FUNDAMENTALS OF COMPUTER AIDED ANALYSIS AND DESIGN (CAA/CAD) OF INTEGRATED CIRCUITS, PROCESSES AND DEVICES

A SELF INSTRUCTIONAL APPROACH

Andres Fortino, Ph.D.

Reston Publishing Company, Inc., Reston, Virginia
A Prentice-Hall Company

ISBN: 0-8359-2120-4

Reston Publishing Company, Inc.
A Prentice-Hall Company
Reston, Virginia 22090

10 9 8 7 6 5 4 3 2 1

Printed in the United States of America.

An offering to the Mother of all life.

TABLE OF CONTENTS

INTRODUCTION

This workbook is intended for those students of semiconductor technology who wish to develop understanding and mastery of the technology of integrated circuits, devices and processes. In the twelve cases studied, all aspects of the technology are covered: process analysis and design, device analysis and characterization, and the rudiments of computer aided circuit analysis. In all cases, the aim is to have the student understand and manipulate all the elements of the analysis of devices and processes. He is expected to do computer work throughout this workbook with each case study culminating in a student-generated computer program to solve the given problems. This necessitates a computer laboratory portion of the course if the course is to be taught didactically. The workbook is also designed for those practicing in the field to be able to work through each case on their own.

In the author's experience, all assigned computer programs are of practical importance even beyond the course work. The circuit analysis case is treated simply using a two transistor circuit model. Industry programs, such as SPICE for circuit analysis, may also be successfully used. The use of industry programs is referenced in each case where appropriate.

The objectives of the workbook are:

1. For the student to develop an understanding of the basic aspects of integrated circuit technology, solid state science and analysis of devices, processes and circuits;
2. For the student to develop and exercise the very rudimentary analysis tools in the form of computer programs necessary to work with the technology;
3. For the student to develop a technical expertise necessary to perform characterization of devices, circuits and processes, and eventually to design with the tools developed.

Organizationally the workbook is designed to guide the students from the easier, basic cases to the more complex. Since it is cumulative, students approaching this material for the first time should begin with Case 1 and work forward. More advanced students, with the appropriate background, may work through one or several cases in an area independently to develop skill in that aspect of the technology. A series of experiments in the characterization of MOSFET devices complementing the text are included in Appendix C.

This workbook may also be used as a source of exercises to accompany a course in integrated circuits and devices. One may organize a complete course around the workbook with supplementary notes and a number of textbooks as references. On the other hand a particular text may be followed using the workbook as a source of exercises and computer problems.

The author wishes to acknowledge the help of many students who patiently worked out the problems in their draft form. Of special note was the assistance of Mr. John Helferty and Mr. Dov Vider. I am also indebted to Mrs. Oksana Bilyk who was instrumental in the preparation of the manuscript.

RESOURCE BIBLIOGRAPHY

1. W.N. Carr and J.P. Mize, MOS/LSI Design and Application, McGraw-Hill, N.Y., 1972.

2. R.A. Colclaser, Microlelectronics: Processing and Device Design, John Wiley and Sons, N.Y., 1980.

3. G. Dearnaley, J. H. Freeman, R. S. Nelson and J. Stephen, Ion Implantation, Defects in Crystalline Solids Series, Vol. 8, North-Holland Publishing Co., Amsterdam, 1973.

4. E. Douglass and A. Dingwall, "Ion Implantation for Threshold Control of COSMOS Circuits", IEEE; TED-21, No. 6, p. 324-331, (1974).

5. P. E. Gise and R. Blanchard, Semiconductor and Integrated Circuit Fabrication Techniques, Reston, Reston, Va., 1979.

6. A. B. Glaser and G. E. Subak-Sharpe, Integrated Circuit Engineering, Addison-Wesley, Reading, Mass., 1979.

7. A. Goetzberger, "Ideal MOS Curves for Silicon", Bell System Technical Journal, 45, 1097 (1966).

8. A. S. Grove, Physics and Technology of Semiconductor Devices, Wiley and Sons, N.Y., 1967.

9. J.C. Irvin, "Resistivity of Bulk Silicon and Diffused Layers in Silicon", Bell System Technical Journal, 41, p.387 (1962).

10. W. S. Johnson and J. F. Gibbons, Projected Range Statistics in Semiconductors, Stanford University Press, 1969.

11. A.G. Milnes, Semiconductor Devices and Integrated Electronics, Van Nostrand Reinhold, N.Y., 1980.

12. R.S. Muller, and T.I. Kamins, Device Electronics for Integrated Circuits, John Wiley, N.Y., 1977.

13. A.B. Phillips, Transistor Engineering, McGraw-Hill, N.Y., 1962.

14. P. Richman, MOS Field-Effect Transistors and Integrated Circuits, Wiley-Interscience, N.Y., 1973.

15. V.L. Rideout, F.H. Gansslen and A. Leblanc, "Device Design Considerations for Ion Implanted n-Channel MOSFETS", IBM Journal of Research and Development, Jan. 1975, pp. 50-59.

16. B.G. Streetman, Solid State Electronic Devices, Prentice-Hall, Englewood Cliffs, N.J., 1980.

17. S. M. Sze, Physics of Semiconductor Devices, Second ed., Wiley-Interscience, N.Y., 1981.

18. W.C. Till and J.T Luxon, Integrated Circuits: Materials, Devices and Fabrication, Prentice-Hall, Englewood Cliffs, N.J., 1982.

19. W. Van Gelden and E.H. Niccolian, "Silicon Impurity Distribution as Revealed by Pulsed MOS CV Measurements", Journal of the Electrochemical Society, Jan. 1971, pp. 138-141.

20. H.F. Wolf, Semiconductors, Wiley-Interscience, N.Y., 1971.

CASE 1: RESISTANCE

Objectives

1. Understand the fundamentals of doped semiconductors, carrier concentration, and mobility.

2. Compute resistivity from given doping levels.

3. Compute the resistance of given cross-sectional area resistors, the current flow and estimate material electrical breakdown.

Laboratory

Develop a computer program that computes the resistance of a given geometry integrated resistor from the given doping levels.

References

1. Grove, Chapter 4.

2. Glaser and Subak-Sharpe, Chapter 4.3.

3. Sze, Chapter 1.5.

4. D.M. Caughey and R.E. Thomas, "Carrier Mobilities in Silicon Empirically Related to Doping and Field", Proceedings of the IE' pp 2192-2193, Dec. 1967.

FUNDAMENTALS OF DOPED SEMICONDUCTORS

A very important characteristic is the intrinsic carrier concentration n_i or the number of electrons in the conduction band (or holes in the valance band) at any given temperature in a pure, undoped semiconductor. For room temperature (300°K) silicon this number is:

$$n_i = 1.45 \times 10^{10} \text{ [cm}^{-3}] \qquad (1.1)$$

For other temperatures one may use the plot given in Figure 1.1.

Exercise 1.1

Assuming the relationship between n_i and temperature (in degrees Kelvin) to be:

$$n_i = C_o \times \exp - (E_g/2K_BT)$$

where: E_g = 1.1 [eV] and
K_B = Boltzman's constant = 8.62×10^{-5} [eV/°K];

find C_o by using the value of n_i at room temperature given in Equation 1.1. Plot the results on semi-log graph paper and compare the results to Figure 1.1.

CARRIER CONCENTRATIONS

When a semiconductor is doped with impurities that have one less valence electron or one more valence electron than the host element, an excess of electrons or of holes above the intrinsic level occurs. For silicon, dopants that increase the hole concentration are valence 3 materials (see the Periodic Chart of Figure 1.2) such as boron and are called p type dopants. Those that increase the electron concentration are valence 5 elements such as arsenic and phosphorus and are called n type dopants.

The excess carrier concentration is found from the fact that in thermal equilibrium the following must be satisfied:

$$p \times n = n_i^2 \qquad (1.3)$$

and also charge neutrality must be satisfied:

$$p - n + N_A - N_D = 0; \qquad (1.4)$$

where: p = hole concentration;

n = electron concentration;

N_A = acceptor (p type) impurity concentration;

N_D = donor (n type) impurity concentration.

From all this we obtain:

for n type Si	for p type Si
$n \cong N_D$	$p \cong N_A$
$p \cong n_i^2/N_D$	$n \cong n_i^2/N_A$

Exercise 1.2

Fill out the following table:

Dopant [cm-3]	Impurity type	Material type	Carrier concentration p	n
10^{14} B				
2×10^{17} As				
5×10^{20} P				

MOBILITY

The mobility of each type of carrier changes considerably as a function of doping ion concentration. It is a function of the total impurity concentration, even for a compensated semiconductor. The bulk mobility for silicon (different from surface mobility) is plotted in Figure 1.3. We see that as we approach the intrinsic level (low doping: 10^{14}[cm-3]) the mobility is nearly twice what it is at high doping levels (10^{20}[cm-3]). Also the mobility for electrons is nearly three times that of holes, a very important observation as we shall see in later case studies.
These two curves have been approximated by closed form expressions given below:

For holes:

$$\mu_p = \frac{495. - 47.7}{1+(C_T/6.3\times10^{16})^{0.72}} + 47.7 \text{ [cm}^2\text{/V sec]}; \qquad (1.5)$$

and for electrons:

$$\mu_n = \frac{1330.0 - 65.0}{1 + (C_T/8.5\times10^{16})^{0.76}} + 65 \ [cm^2/V\ sec]; \qquad (1.6)$$

where C_T is the sum of all impurity ions. Normally,one uses the curves in Figure 1.3 which is sufficient for practical purposes; the equations above on the other hand become very useful for computer implementation.

Exercise 1.3

Compute the mobilities of electrons and holes in silicon using the equations above at the following doping levels and compare to that given in the graph of Figure 1.3.

Doping	Electron mobility		Hole mobility	
[cm-3]	equation	graph	equation	graph
10^{14}				
10^{15}				
10^{16}				
10^{17}				
10^{18}				
10^{19}				
10^{20}				

The value of mobility for intrinsic material and many other useful constants for important semiconductors are given in Table 1.1. We shall have occasion to refer to this table very often in future cases.

RESISTANCE

The resistance of a volume of material to the flow of current is given by:

$$R = \rho L/A \ \ [ohms] \qquad (1.7)$$

where ρ is the resistivity, L the length of the material and A its cross-sectional area (assuming it to be constant throughout the length of the device).

The resistivity is the reciprocal of conductivity:

$$\rho = 1/\sigma \text{ [ohm cm]}, \quad (1.8)$$

and conductivity is written as:

$$\sigma = q \mu n \text{ [mhos/cm]}. \quad (1.9)$$

where: $q = 1.6 \times 10^{-19}$ [coulombs];

μ = mobility of the carriers [cm^2/V sec]; and

n = number of carriers [cm^{-3}].

For metals, ρ is single valued at a given temperature (see Table 1.2 for typical values) and the resistance of such a conductor may be computed directly using Equation 1.7.

Exercise 1.4

An aluminum conductive line which connects 528 gates in an integrated circuit is 10,000Å thick, 10 microns wide and .1250 [cm] long. Find its resistance.

Repeat this problem using gold in place of Al and compare the results.

For a semiconductor where there are two types of carriers (electrons and holes), conductivity is expressed as:

$$\sigma = q \mu_n n + q \mu_p p \text{ [mhos/cm]} \quad (1.10)$$

Measurements for silicon produced the graph in Figure 1.4.

Exercise 1.5

Compute the resitance of an arsenic diffused line in a 10^{15} [cm^{-3}] p Si substrate. The device cross-section is shown below:

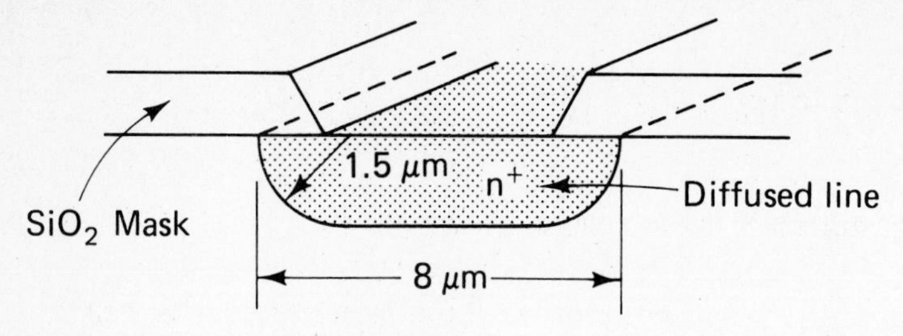

Figure 1.5

Assume that the sidewalls may be approximated by a cylinder of indicated radius. Assume a 1 [cm] length and the doping levels given below. Compute the resistance using the value of resistivity from the graph as well as that computed using Equation 1.10 and compare.

Arsenic doping [cm^{-3}]	resistivity [ohm cm] from graph	resistivity [ohm cm] from equation	resistance [ohms] using graph	resistance [ohms] using equation
10^{15}				
10^{17}				
10^{20}				

Exercise 1.6

Compute the resistance of a polysilicon conductive line connecting several gates together in a 16K RAM. It is 125 microns long, 1 micron thick and 25 microns wide and doped with 10^{19} [cm^{-3}] boron ions. The substrate is 2 [ohm cm] n Si.

OHM'S LAW

Once resistance has been found as above, we shall use the simple relationship between current and voltage:

$$V = I\,R \text{ [volts]}. \qquad (1.11)$$

Another useful way of expressing this relationship is:

$$J = \sigma E \text{ [volts/cm}^2\text{]} \qquad (1.12)$$

where: J = current density = I/A [amps/cm^2] and

E = electric field = V/L [volts/cm].

BREAKDOWN

The voltage across a device may be increased until an certain electric field level is reached at which point the device breaks down and excessively large currents flow. This is known as breakdown and the limiting fields for various semiconductor materials are given in Table 1.1.

Exercise 1.7

The region between source and drain at the surface of a MOSFET, the channel, may be considered a resistor. At a particular gate voltage, there are 10^{18} [electrons/cm^3] in this region, and the thickness of the layer may be approximated to be 500Å deep. If the device length is 20 microns and its width is 50 microns, find the current that flows when the source to drain voltage is 10 [volts]. The substrate is 2 [ohm cm] p Si.

What is the lateral (source to drain) electric field?

Is the semiconductor in the channel under breakdown?

If the source to drain voltage is now increased to 100 [volts], what is the electric field? Does breakdown occur?

Computer Laboratory

In preparation for using the above formulas in later programs, we shall exercise and codify them in this laboratory.

A. Write a computer program that will only accept temperature and doping densities of either acceptors or donors or both superimposed (called compensation). The program should compute and print:

   a. n_i;
   b. mobilities;
   c. carrier concentrations; and
   d. conductivity and resistivity.

B. This program should also be able to compute the resistance of a semiconductor volume given the physical dimensions and the doping and temperature as above. It should also compute and print the electric field and current when a given voltage is applied.

C. Using this code, rework problems 1.2, 1.3, 1.5, 1.6, and 1.7, to validate the computer code.

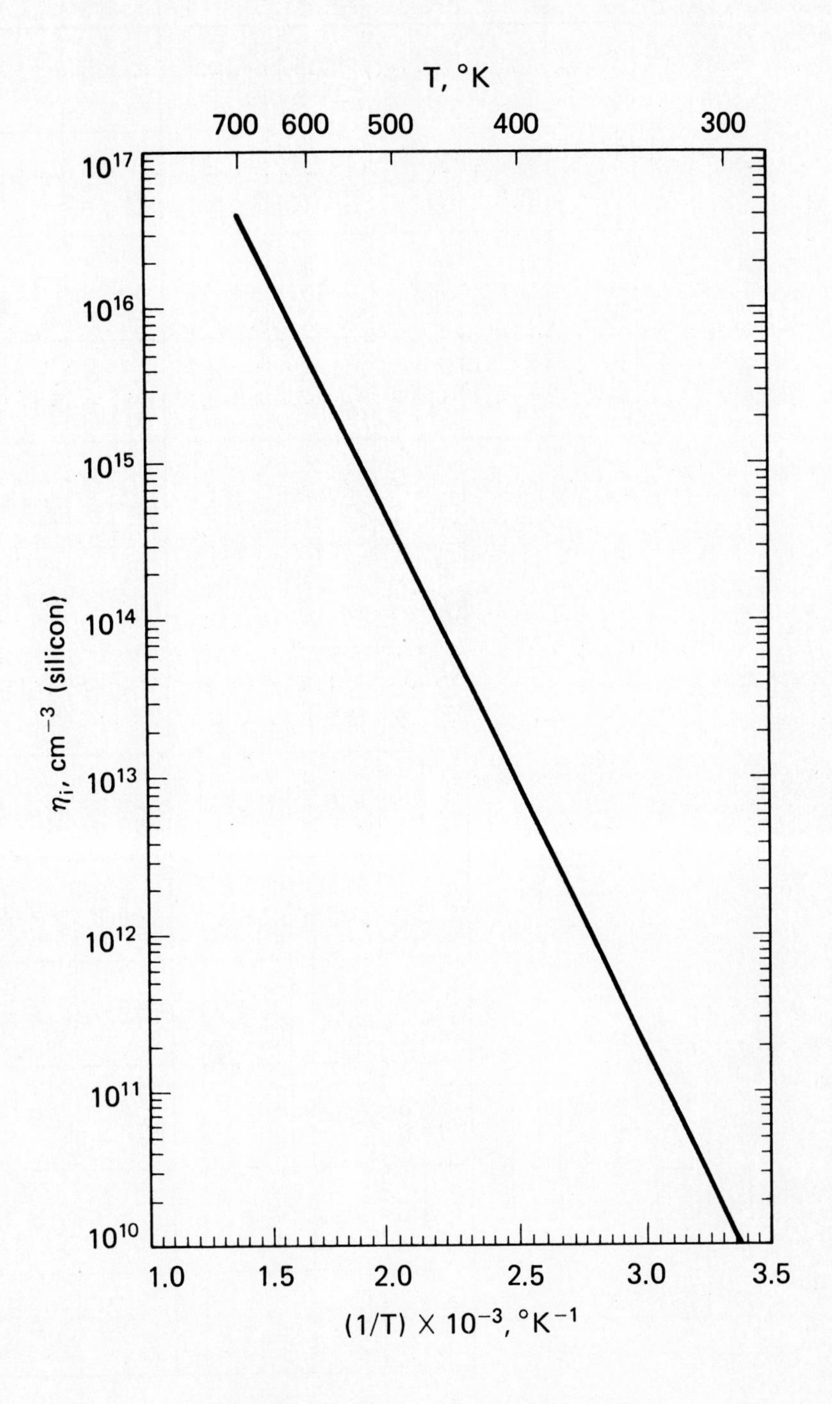

Figure 1.1 - Intrinsic carrier concentration of silicon (n_i) as a function of temperature.

Legend . . . Atomic number | Density (G/cm³)
Element
Atomic mass
u

Period or principal quantum number	IA	IIA	IIIA	IVA	VA	VIA	VIIA	VIIIA			IB	IIB	IIIB	IVB	VB	VIB	VIIB	VIIIB
1	1 H 1.0080																	2 1.85 He 4.0026
2	3 0.53 Li 6.939	4 Be 9.012											5 2.34 B 10.81	6 2.62 C 12.011	7 N 14.09	8 O 15.999	9 F 18.998	10 Ne 20.18
3	11 0.97 Na 22.99	12 1.74 Mg 24.31											13 2.70 Al 26.98	14 2.33 Si 28.09	15 1.82 P 30.98	16 2.07 S 32.06	17 Cl 35.46	18 Ar 39.95
4	19 0.86 K 39.10	20 1.55 Ca 40.08	21 3.00 Sc 44.96	22 4.50 Ti 47.90	23 5.80 V 50.95	24 7.19 Cr 52.01	25 7.43 Mn 54.94	26 7.86 Fe 55.85	27 8.90 Co 58.93	28 8.90 Ni 58.71	29 8.96 Cu 63.54	30 7.14 Zn 65.38	31 5.91 Ga 69.72	32 5.32 Ge 72.59	33 5.72 As 74.91	34 4.08 Se 78.96	35 3.12 Br 79.92	36 Kr 83.80
5	37 1.53 Rb 85.48	38 2.6 Sr 87.63	39 4.5 Y 88.92	40 6.49 Zr 91.22	41 8.55 Nb 92.91	42 10.2 Mo 95.95	43 11.5 Tc	44 12.2 Ru 101.1	45 12.4 Rh 102.91	46 12.0 Pd 106.4	47 10.5 Ag 107.88	48 8.65 Cd 112.41	49 7.31 In 114.82	50 7.30 Sn 118.70	51 6.68 Sb 121.76	52 6.24 Te 127.61	53 4.92 I 126.91	54 Xe 131.30
6	55 1.87 Cs 132.91	56 3.50 Ba 137.36	57 6.70 La* 138.92	72 13.1 Hf 178.50	73 16.6 Ta 180.95	74 19.3 W 183.86	75 21.0 Re 186.22	76 22.4 Os 190.2	77 22.5 Ir 192.2	78 21.4 Pt 195.09	79 19.3 Au 197.0	80 13.53 Hg 200.61	81 11.85 Tl 204.39	82 11.24 Pb 207.21	83 9.8 Bi 209.00	84 9.4 Po (209)	85 At (210)	86 Rn (222)
7	87 Fr (223)	88 5.00 Ra 226.05	89 10.07 Ac** (227)															
	*Lanthenum series			58 Ce 140.13	59 Pr 140.92	60 Nd 144.27	61 Pm	62 Sm 150.35	63 Eu 152.0	64 Gd 157.76	65 Tb 158.93	66 Dy 162.51	67 Ho 164.94	68 Er 167.27	69 Tm 168.94	70 Yb 173.04	71 Lu 174.98	
	**Actinum series			90 Th 232.04	91 Pa 231.04	92 U 238.07	93 Np 237.05	94 Pu (244)	95 Am (243)	96 Cm (247)	97 Bk (247)	98 Cf (251)	99 Es (254)	100 Fm (257)	101 Md (258)	102 No (259)	103 Lw (260)	
	104 (260)	105 (260)																

Figure 1.2 - Periodic chart of the elements showing atomic number and atomic mass.

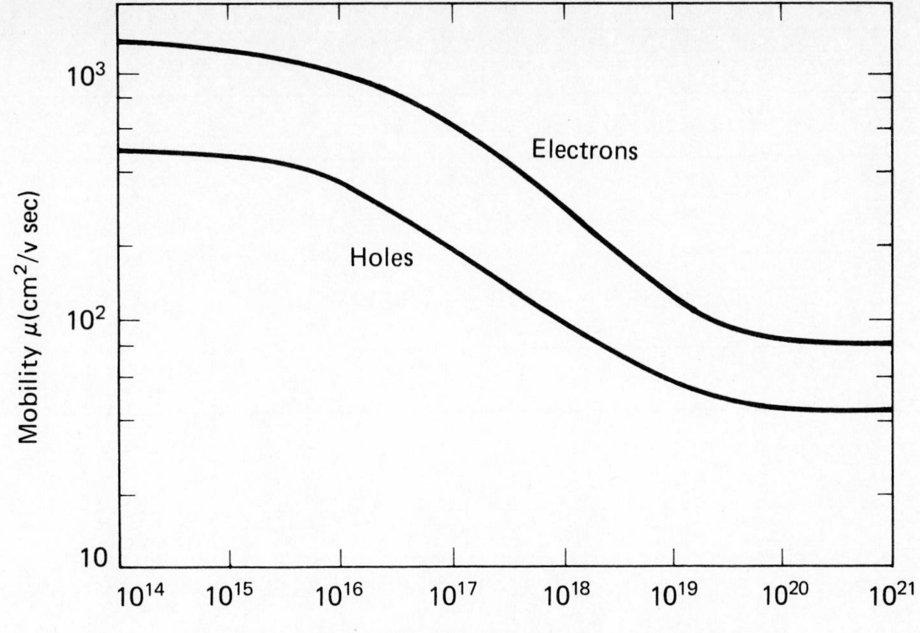

Figure 1.3 - Mobility of electrons and holes for silicon at room temperature as a function of total impurity concentration. (After Conwell. Reprinted with permission from Proceedings of the IRE, Vol. 46, pp.1281-1300, 1958.)

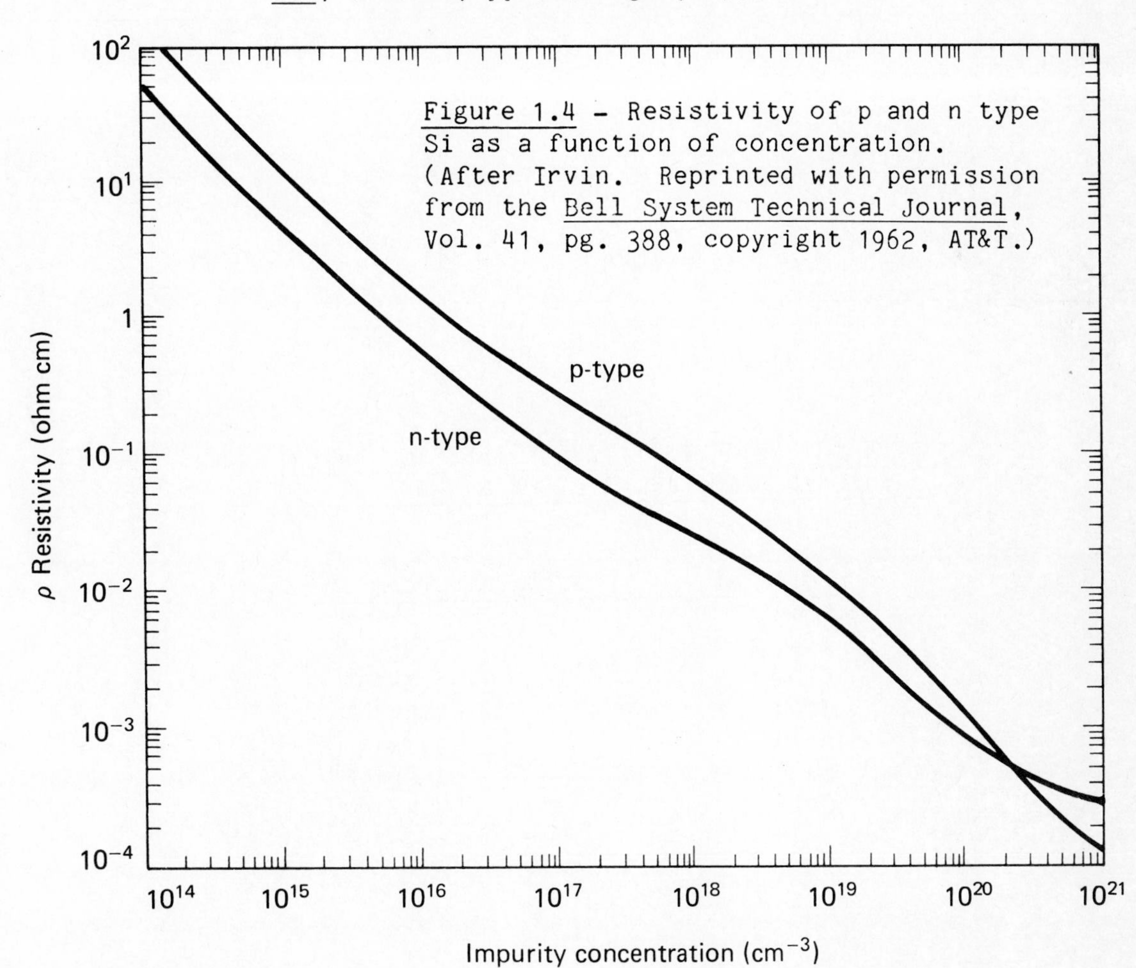

Figure 1.4 - Resistivity of p and n type Si as a function of concentration. (After Irvin. Reprinted with permission from the Bell System Technical Journal, Vol. 41, pg. 388, copyright 1962, AT&T.)

TABLE 1.1 - IMPORTANT PROPERTIES OF GERMANIUM, SILICON, GALLIUM ARSENIDE SILICON DIOXIDE AND SILICON NITRIDE AT 300°K.

Property	Ge	Si	GaAs	SiO_2	Si_3N_4
Atomic weight (or molecular)	72.6	28.09	144.63	60.08	140.28
Atoms per cm^3	4.42×10^{22}	5×10^{22}	2.21×10^{22}	2.3×10^{22}	1.28×10^{22}
Density [g/cm^3]	5.32	2.33	5.32	2.27	3.0
Energy gap [eV]	0.67	1.11	1.4	8.0	5.0
Intrinsic carrier concentration n_i [cm^{-3}]	2.4×10^{13}	1.45×10^{10}	9.6×10^{6}	-	-
Intrinsic mobilities [cm^2/V sec]					
electrons	3900	1350	8600	Insulator	Insulator
holes	1900	480	250		
Relative permittivity	16.3	11.7	12.0	3.9	6 - 9
Breakdown field [V/μm]	8	30	35	600	500
Melting point [°C]	937	1450	1238	1700	1900
Thermal conductivity [watt/cm°C]	0.6	1.5	0.81	0.014	-

TABLE 1.2
ELECTRICAL RESISTIVITY OF SELECTED MATERIALS SILICON DIOXIDE AND SILICON NITRIDE AT 300°K.

Material	Resistivity [ohm cm]
Aluminum	2.9×10^{-6}
Copper	1.7×10^{-6}
Iron	9.8×10^{-6}
Lead	2.1×10^{-5}
Silver	1.8×10^{-6}
Gold	2.4×10^{-6}

CASE 2: RESISTANCE

Objectives

1. Understand the concept of sheet rho and ohms per square.

2. Compute the sheet rho of ion implanted and uniformly doped layers.

3. Understand the theory and practice of four point probe measurements.

Laboratory

Develop a computer program that computes the resistance of diffused layers and computes sheet rho from given parameters of epitaxial and ion implanted layers.

References

1. Glaser and Subak-Sharpe - Chapters 4.1 and 4.2 and 11.2.

2. J.C. Irvin, "Resistivity of Bulk Silicon and Diffused Layers in Silicon", Bell System Tech. Journal, No. 41, page 387, 1962.

SHEET RESISTANCE

For doping concentration profiles that form either a junction or have a many-orders-of-magnitude doping step, the resistance measured on the wafer may be given by:

$$R = \rho L/A \quad [\text{ohms}] \qquad (2.1)$$

It may be expressed as:

$$R = R_S L/W = R_S n \quad [\text{ohms}] \qquad (2.2)$$

where the sheet rho $R_S = \rho/t$, t is the layer thickness and n is the aspect ratio. When the layer is non-uniform, an average ρ is commonly employed. The aspect ratio is found by dividing the surface area into square (arbitrary size) where l is the number of squares of length and w the number of squares in width.

Example 2.1

Consider an n type expitaxial layer on 2 [ohm cm] p type silicon. It is doped with a phosphorus concentration of 10^{18} [cm^{-3}] and assumed it to be uniform for a depth of 10 [μm] forming a rectangle 500 [μm] long by 10 [μm] wide.

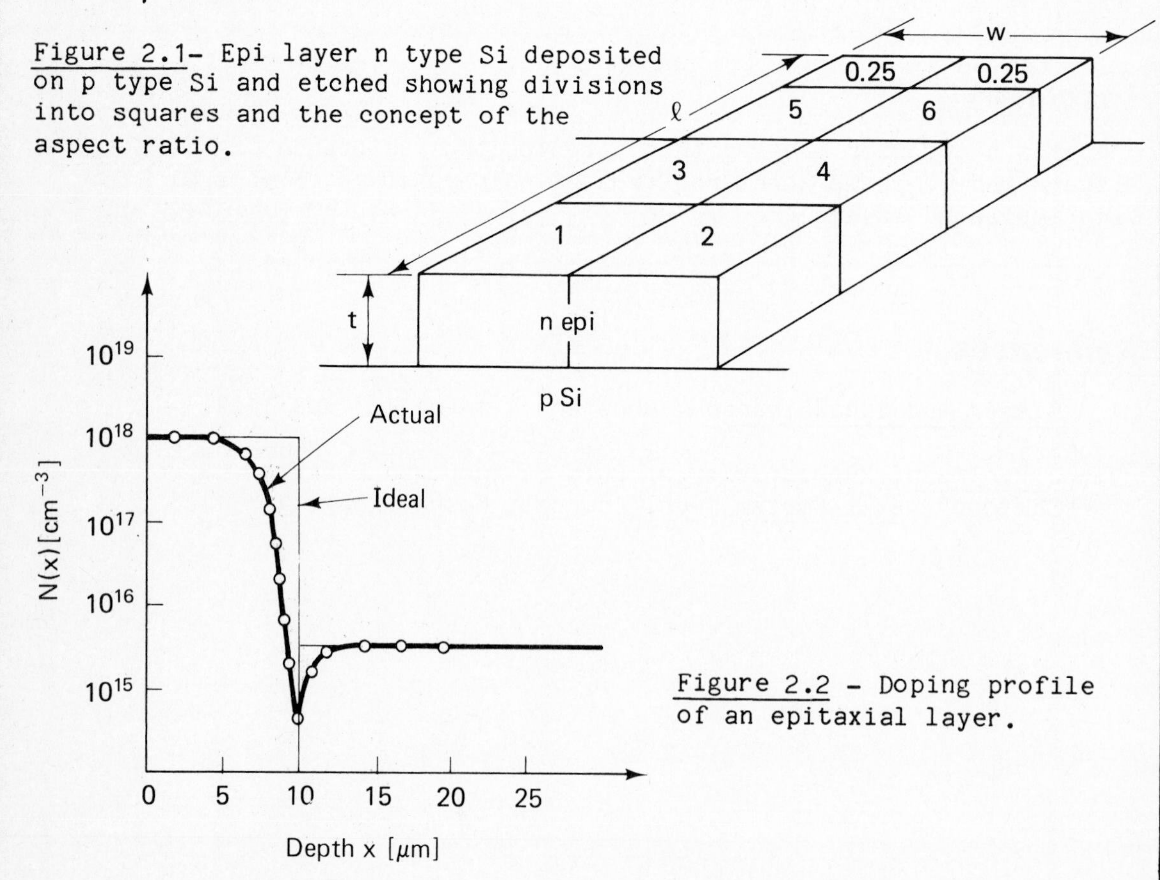

Figure 2.1- Epi layer n type Si deposited on p type Si and etched showing divisions into squares and the concept of the aspect ratio.

Figure 2.2 - Doping profile of an epitaxial layer.

a. What is the background doping?

Answer: For 2 [ohm cm] p type we see in Figure 1.4 that $N_A = 6 \times 10^{15}$ [cm^{-3}].

b. What is the sheet rho of the epitaxial layer?

Answer: For 10^{18} [cm^{-3}] phosphorous doped Si, ρ = .025 [ohm cm] from Figure 1.4; and $R_S = \rho/t$ = .025 [ohm cm]/10×10^{-4} [cm].

$$R_S = 25 \text{ [ohms/square]}$$

c. What is the aspect ratio (n = L/W) for this device?

Answer: n = 500 [μm]/10 [μm] = 50 squares.

d. What is the resistance of this resistor?

Answer: $R = R_s$ n = 25 [ohms/square] x 50 squares = 1.25 [Kohms].

Exercise 2.1

Consider a "uniformly" doped diffusion of boron into n type Si. The diffusion is doped with an impurity concentration of 10^{20} [cm^{-2}] and a junction depth of 5 [μm]; the substrate is 2 [ohm cm] p Si.

A. Draw in the given semi-log graph an approximate profile distribution. Mark the junction.

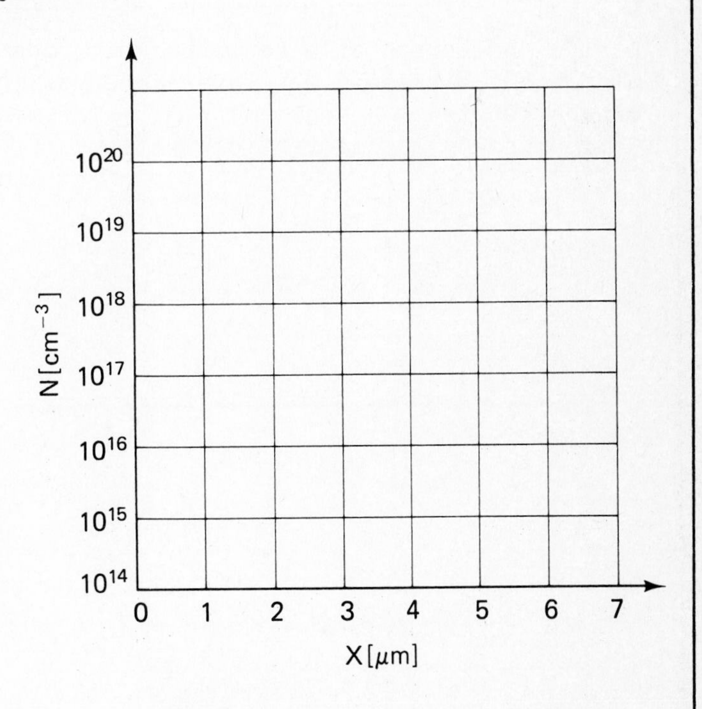

Figure 2.3 - Plot of a uniform diffusion profile.

B. Compute the sheet rho of the diffusion.

C. An interconnection between two devices made with this diffusion is 125 microns long and 25 microns wide. Compute the number of squares and the resistance of this interconnection.

Exercise 2.2

Design the length of a resistor which must be 100 [Kohm], from a 15 micron thick arsenic epi layer doped to 10^{18} [cm^{-3}] arsenic concentration on 2 [ohm cm] p Si. The minimum printable line width is 5 microns and assume vertical walls.

NONUNIFORM PROFILES

Profiles which are nonuniformly doped are entirely more common than uniform ones. In this case, to achieve a greater accuracy in building resistors and for analysis of other devices, an integration is required:

$$R_S = \rho_{s/t} = \frac{1}{\int_0^t q\, \mu(N) N(x) dx} \text{ [ohms]} \qquad (2.3)$$

where $\mu(N)$ is the mobility of the majority carriers as a function of doping, N, which itself is varying with depth x.

We shall implement this equation in a computer algorithm described below. Because of the usual many-orders-of-magnitude change in doping in a short space, a simple predictor-corrector scheme will be used for integration. Consider the arbitrary doping profile shown in Figure 2.4. The procedure is given below:

1. Decide on the number of integration points desired, and discretize N(x) into an array of that many points.

EXAMPLE: let us say we chose S points:

N(x) 0<x<t analytical ⟹ DOP (I) 1<I<S discrete

FORTRAN CODE:

```
      DX = T/(S+1)
      X = 0.0
      DO 10 I = 1,S
      DOP(I) = N(X)
   10 X = X+DX
```

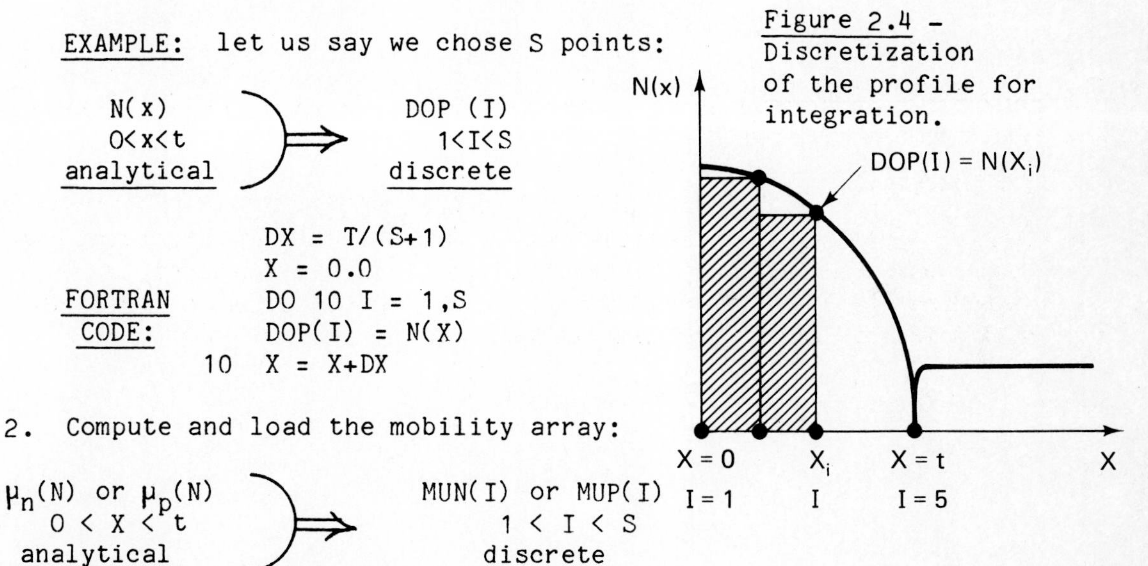

Figure 2.4 - Discretization of the profile for integration.

2. Compute and load the mobility array:

$\mu_n(N)$ or $\mu_p(N)$ 0 < X < t analytical ⟹ MUN(I) or MUP(I) 1 < I < S discrete

see equations 1.5 and 1.6

3. Decide whether the layer is n or p by the doping type; integrate using the trapezoidal method (assume n type for this example):

analytic expression: Integral = $\int \mu(N)\ N(X)\ dx$

FORTRAN equivalent: INT = INT + (MUN(I)*DOP(I) + MUN(I-1)*DOP(I-1)*DX/2

FORTRAN CODE:

```
      INT = 0
      DO 20 I = 2,S
   20 INT = INT+(MUN(I)*DOP(I)+MUN(I-1)*DOP(I-1))*DX/2
      RS = 1/(Q*INT)
```

Exercise 2.3

Write the computer code for performing the above integration. (Assume N(X) to be constant for this problem.)

Set constants

```
Q = 1.6x10^-19 [coulombs]
T = depth in [cm]
S = number of points
DX = T/(S+1)
```

Discretize the doping array

Compute and load the mobility array

Decide n or p type

Integrate

Compute the sheet rho

Gaussian doping profiles may be expressed analytically as:

$$N(x) = P_o \exp -[(x-\mu)^2/2\sigma^2] + N_B \; [cm^{-3}] \qquad (2.4)$$

N_B = background doping.

P_o = concentration above background $[cm^{-3}]$.

μ = range of profile; distance between surface (x=0) and peak of profile [cm].

σ = standard deviation; distance between peak of profile and .606 of peak of profile [cm].

x = depth into Si [cm].

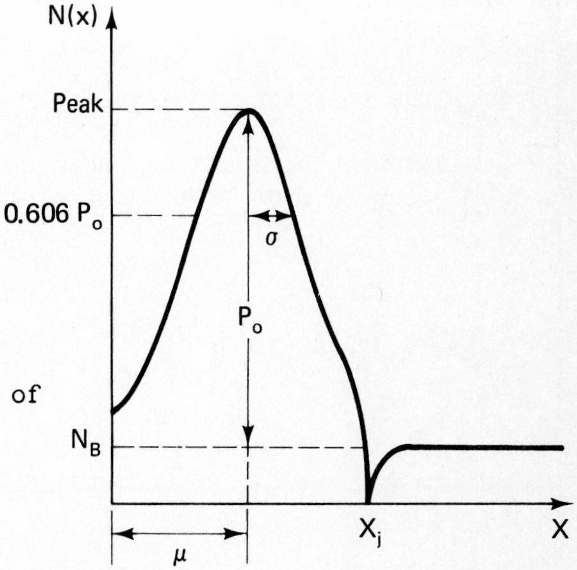

Figure 2.5 - The junction of a gaussian profile.

If the doping of the gaussian layer is of opposite type from background (bulk), then a junction is formed with junction depth given by (see the figure above):

$$x_j = \sqrt{2\sigma^2 \ln(P_o/N_B)} + \mu \; [cm] \qquad (2.5)$$

Exercise 2.4

A gaussian profile has the following parameters:

Peak concentration (P_O) = $5x10^{16}$ [cm^{-3}] (n type);

Range (μ) = 0 [cm];

Standard deviation (σ) = .5 [µm] = $.5x10^{-4}$ [cm];

Background doping (N_B) = $-1x10^{15}$ [cm^{-3}](p type).

a. Enter these parameters into Equation 2.4 and fill in the following table:

x [µm]	0.0	0.1	0.2	0.3	0.4	0.5	0.6	0.7	0.8
N(x)[cm^{-3}] x10^{16}									

x [um]	0.9	1.0	1.1	1.2	1.3	1.4	1.5	1.6	1.7
N(x)[cm^{-3}] x10^{16}									

b. Plot this curve using linear graph paper.

c. Compute the junction depth using Equation 2.5 and compare to that observed from graph just plotted.

Computer Laboratory

Using the computer program written in Exercise 2.3 as a base, code and exercise a computer program to:

1. Compute and print the sheet rho of a uniform doped layer. Use the data and solution of Exercises 2.1 and 2.2 as the two examples to validate the program. Compare your results to those worked out by hand in both cases.

2. Change the program (and save as a new version) so that it computes the sheet rho of a gaussian junction. Code Equation 2.4 as the profile and enter the appropriate costants. Use the data of Exercise 2.4 as the validating example. Make sure to print out the doping array after discretizing the gaussian profile as a check.

FOUR POINT PROBE MEASUREMENTS

Sheet rho measurements may be made using a four point probe arrangement. Figure 2.6 depicts the circuit and probe placement. Again, assuming a constant concentration for the layer being probed, the resistivity is given by:

$$\rho = 2\pi\, a\, V_R/I \quad [\text{ohm cm}] \qquad (2.6)$$

where a is the probe spacing. The sheet rho is given by:

$$R_S = \rho\ /t = (\pi/\ln 2)(V_R/I) \quad [\text{ohm cm}^2] \qquad (2.7)$$

when the measurement is done away from the wafer edges and assuming the layer to be an infinite wide and long sheet. Using this last equation we may solve for the film thickness:

$$t = \rho \ln 2/\pi R_G\ V_G/V_R \quad [\text{cm}] \qquad (2.8)$$

where R_G is a standard resistance and V_G is the measured voltage across it (see Figure 2.6). This method is accurate to +/- 10% for film thicknesses at least 1 μm thick. One may use this method of analysis to profile a doped layer, which is called spreading resistance measurements.

Using this method, resistivities in the range of .001 [ohm cm] (10^{19} [cm^{-3}] doping) to 500 [ohm cm] (10^{14} [cm^{-3}] doping) may be measured.

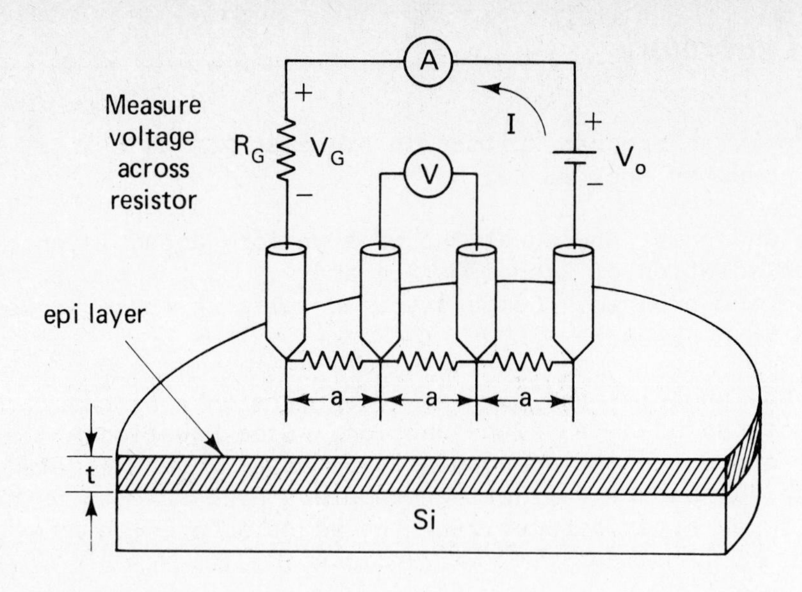

Figure 2.6 - Four point probe arrangement for measuring resistivity of doped layers.

Exercise 2.5

The resistivity of a certain epitaxial layer has been measured by four point probe. The resulting voltage is 10 [V] when a controlled current of 5 [mA] is introduced on the outer probes; (R_G = 1 [Kohm] , V_G = 5 [V]). The probe spacing is 25 [mils] center to center. Find the sheet rho and resistivity of the epi layer.

CASE 3: CAPACITANCE

Objectives

1. Understand the fundamentals of MOS capacitors.
2. Recognize various regions of operation, accumulation, depletion, inversion, and high and low frequency.
3. Analyze capacitance-voltage curves to find V_{FB} and Q_{SS}.
4. Understand measurement techniques.
5. Understand the effect of substrate doping and oxide thickness on capacitance.

Laboratory

Develop a computer code for capacitance-voltage analysis.

References

1. Grove: Chapter 9.
2. Sze: Chapter 9.
3. Richman: Chapter 3.
4. Glaser and Subak-Sharpe: Chapter 3, Section 3.

THE MOS (Metal-Oxide-Silicon) CAPACITOR

The MOS capacitor crossection is shown in Figure 3.1. Corresponding surface charge and potential for the various biasing cases are given in Figure 3.2 for p type Si.

Exercise 3.1

Draw similar diagrams, including the charge density distributions, for an n type substrate for the four cases shown in Figure 3.2.

WORK FUNCTION

Figure 3.3 shows the work function plotted for the metal-silicon system. The work function difference for uniform substrates may be expressed as in Reference 2:

$$\phi_{ms} = \phi_m - (X - E_g/2q - \phi_B) \text{ [volts]} \qquad (3.1)$$

where: ϕ_m = metal work function = 4.35 [V] for Al.

X = semiconductor electron affinity = 5.25 [V] for Si.

E_g = semiconductor band gap [in eV] = -1.2 [eV] for Si.

ϕ_B = bulk fermi level = +/- $(KT/q)\ \ln (N_B/n_i)$ for Si (+ for n type, - for p type).

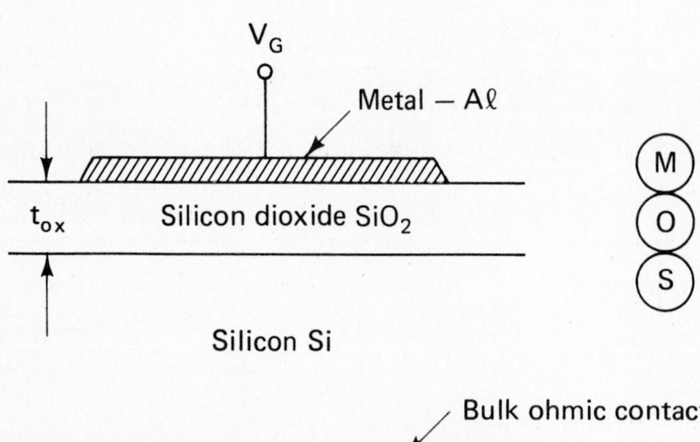

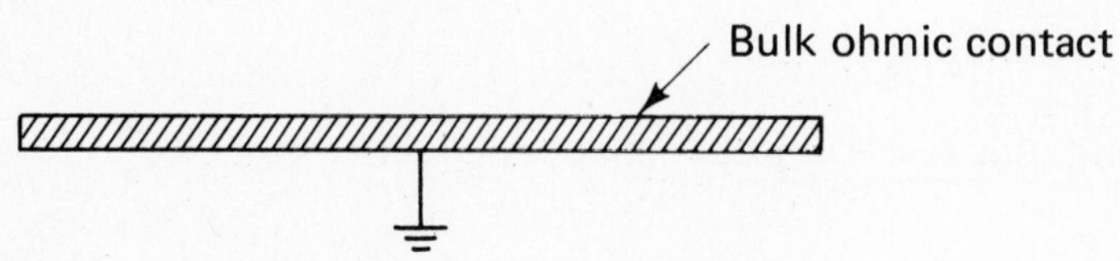

Figure 3.1 - Metal-Oxide-Semiconductor device cross-section

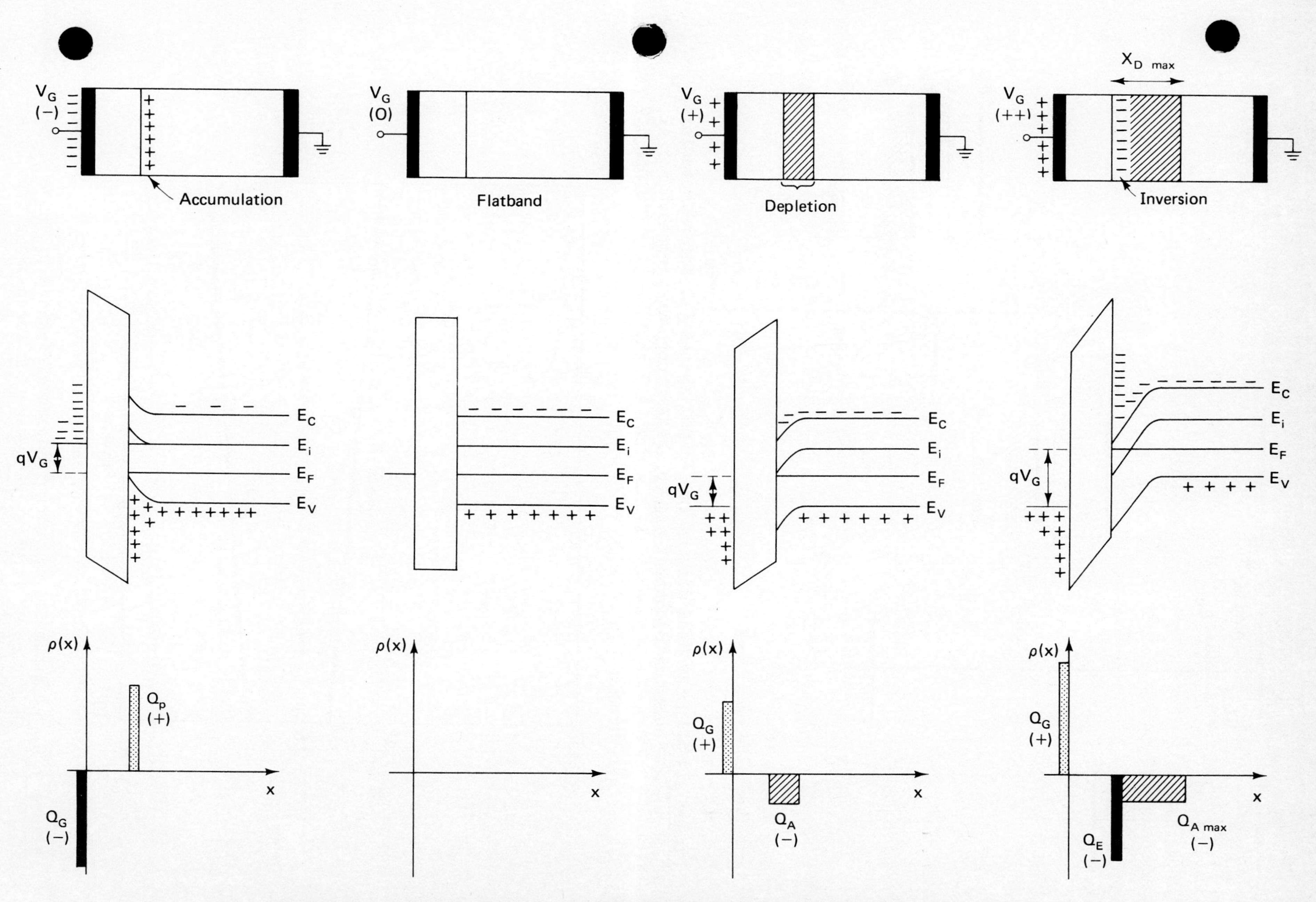

<u>Figure 3.2</u> - MOS capacitor under various bias conditions. Diagrams show charge type and polarity at the surface, surface band bending and charges at the interface.

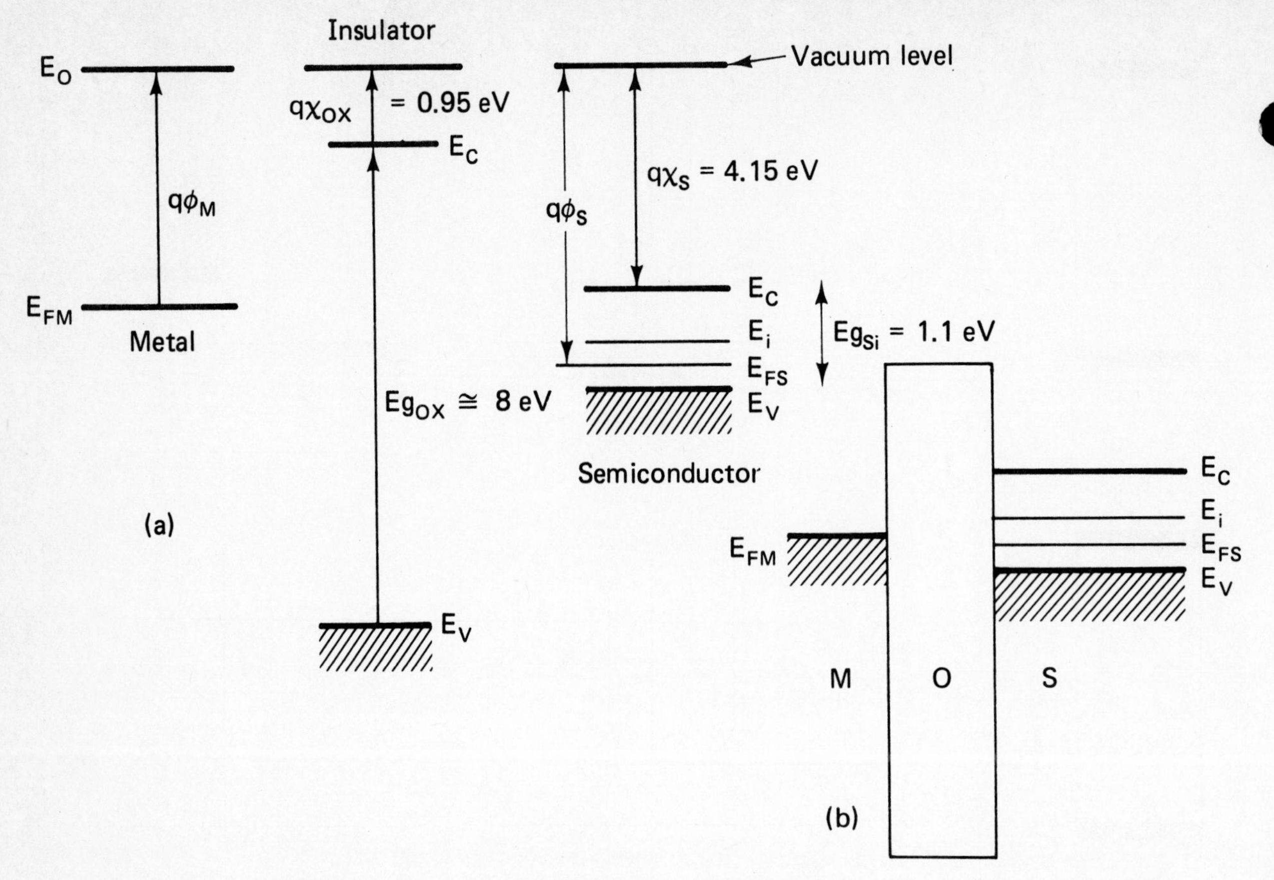

Figure 3.3 - Energy band diagram for an MOS system (a) before materials come into contact and (b) after they have joined. For Al the work function is O_m= 4.1 [eV] and for Si it is O_s= 4.7 [eV].

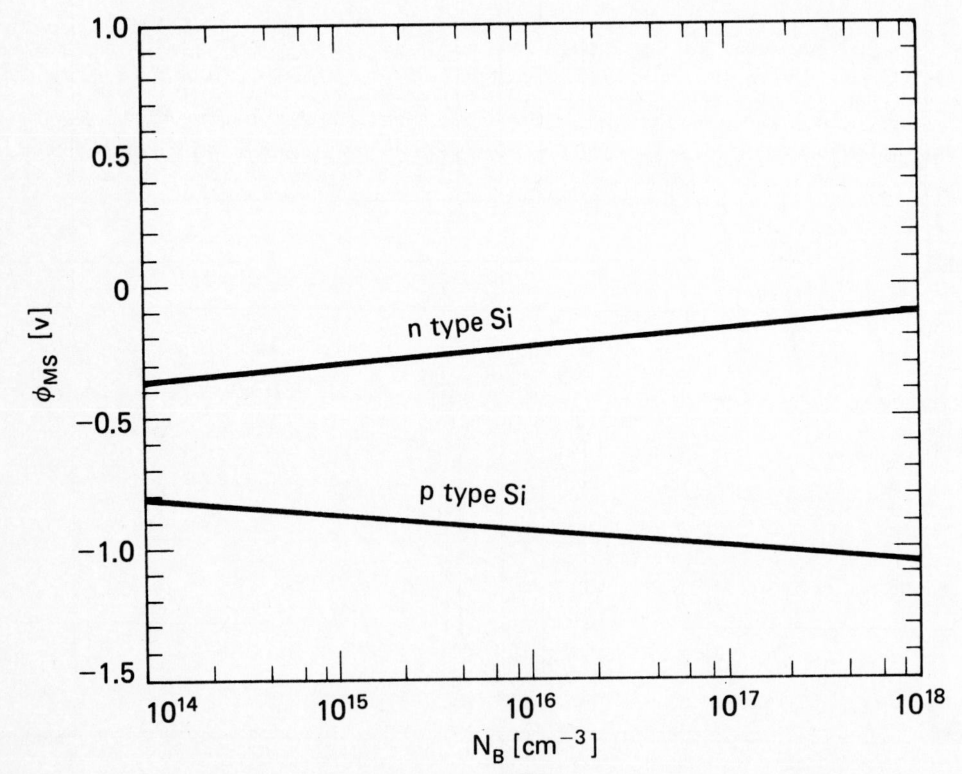

Figure 3.4 - Aluminum-Silicon work function difference ϕ_{ms} as a function of impurity concentration for n and p type Si at 300°C (Reference 11).

Exercise 3.2

Compute ϕ_{ms} using Equation 3.1 for Al-to-10^{15} [cm^3] doped p type Si and Al-to-10^{18} [cm^3] n type Si. Compare to values read from Figure 3.4.

CAPACITANCE-VOLTAGE RELATIONSHIPS

Representative capacitance of the MOS capacitor for various bias conditions is shown in Figure 3.5. Both "low frequency" and "high frequency" curves are shown together with the capacitance model for each case and corresponding bias condition.

Accumulation: $C = C_{ox}\epsilon_o A/t_{ox} = C_{max}$ (3.2)

Depletion: $C = (C_{max} \cdot C_s)/(C_{max} + C_s)$ (3.3)

Silicon capacitance: $C_S = K_S \epsilon_o A/X_D$ (3.4)

Inversion: $C_{min} = (C_{max} \; Cs_{min})/(C_{max} + Cs_{min})$ (3.5)

$Cs_{min} = K_s \epsilon_o A/X_{Dmax}$

$X_{Dmax} = \sqrt{2K_s\epsilon_o(2\,\phi_B) / qN_B}$ (3.6)

The minimum capacitance versus doping at high frequency is plotted in Figure 3.6 as a function of doping and oxide thickness. This is a very useful chart, as we shall see.

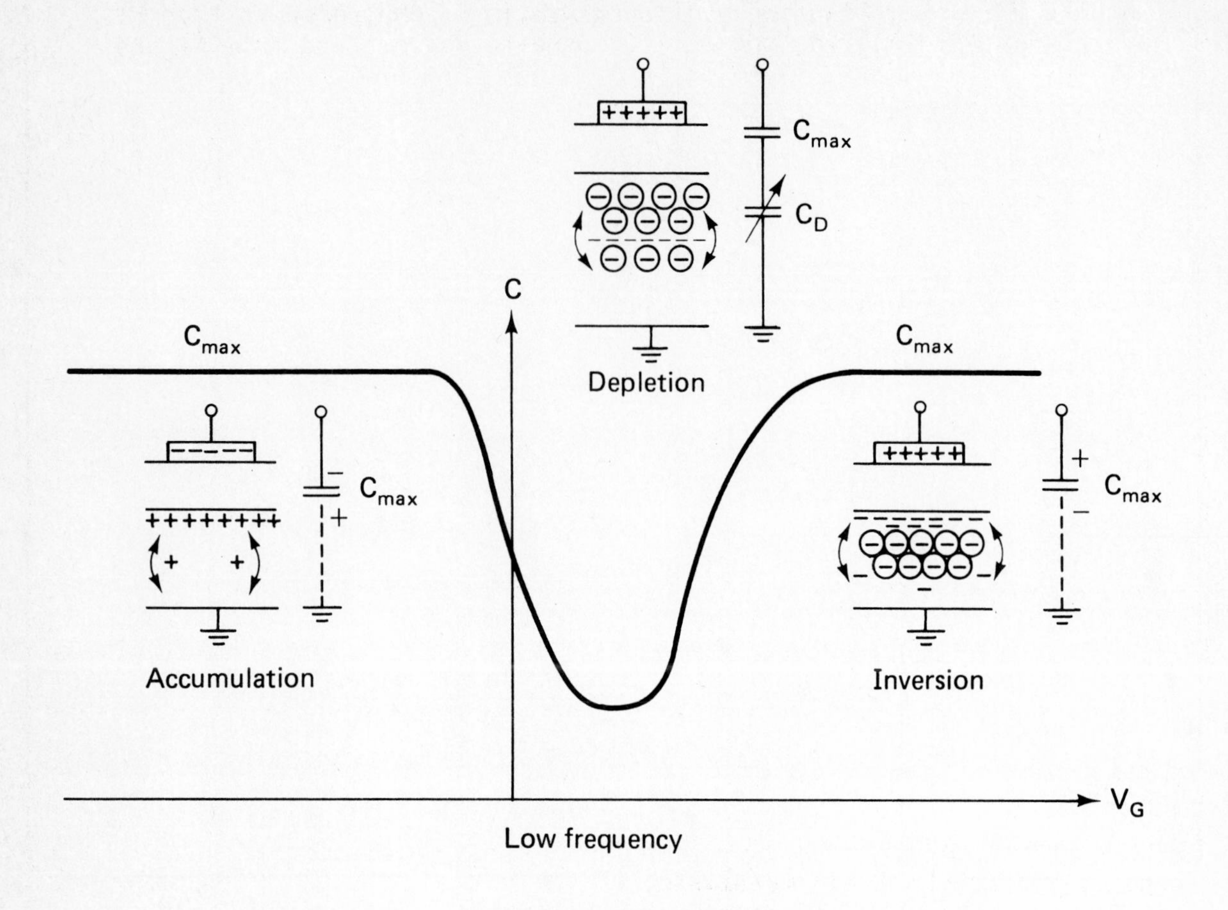

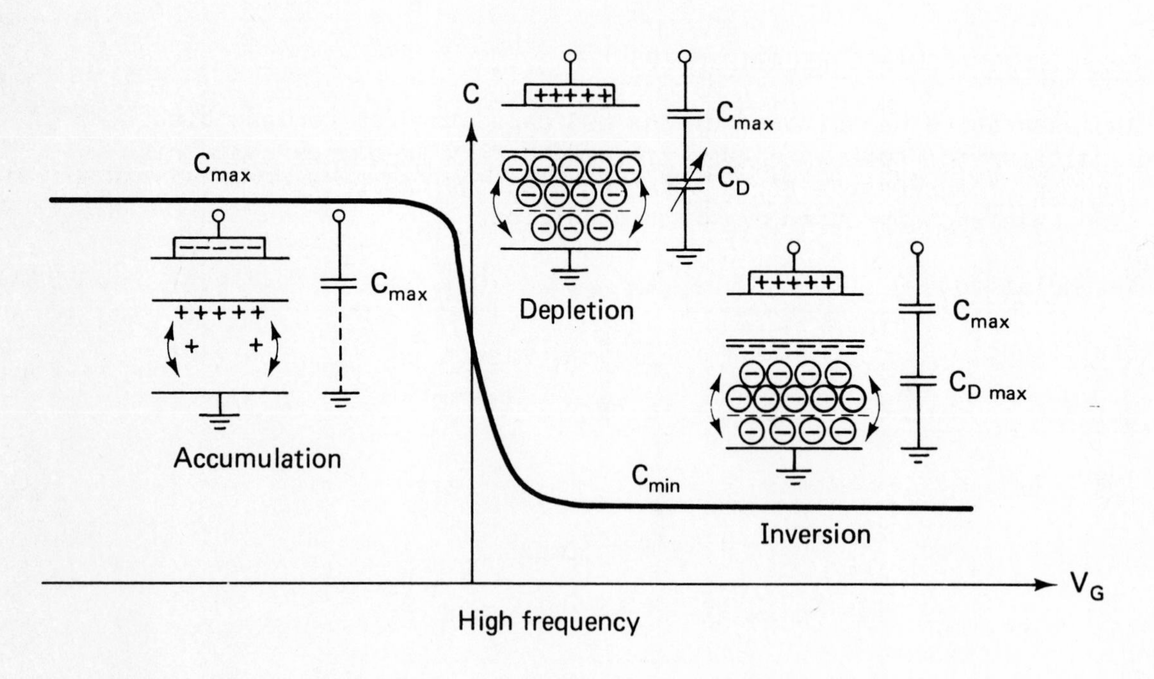

Figure 3.5 - High and low frequency C-V curves for a p type MOS capacitor.

Exercise 3.3

Compute, using Equations 3.5 and 3.6, the inversion capacitance of a 30 [mil] diameter aluminum MOS capacitor, t_{ox} = 1000Å and p type Si doped with 10^{15} [ions/cm^3]. Consult the chart in Figure 3.6 and compare it to your calculation.

FLATBAND

The flatband condition occurs when sufficient gate voltage has been applied to overcome the metal-semiconductor work function difference and any surface potential due to imbedded charges in the insulators:

$$V_{FB} = \Phi_{ms} - [Q_{SS} / (C_{ox}/A)] \text{ [V]}. \qquad (3.7)$$

V_{FB} is defined as the gate voltage necessary to cause the energy bands in the semiconductor to be flat. A more common definition and the one we shall use here is the gate voltage which causes the metal fermi level and the silicon fermi level to line up. Q_{SS} is the charge density assumed to be at the SiO_2 - Si interface.

If there were no work function difference or interface charge, then the capacitance at flatband would be (see Reference 2):

$$C_{FB} = (C_{max} \; C_{SFB}) / (C_{max} + C_{SFB}) \text{ [F]} \qquad (3.8)$$

$$C_{SFB} = 2 K_s \epsilon_o A / L_D \text{ [F]} \qquad (3.9)$$

$$L_D = \text{Debye length} = \sqrt{(2K_s \epsilon_o KT) / (q^2 N_B)} \text{ [cm]} \qquad (3.10)$$

This flatband capacitance is charted in Figure 3.7, another very useful graph.

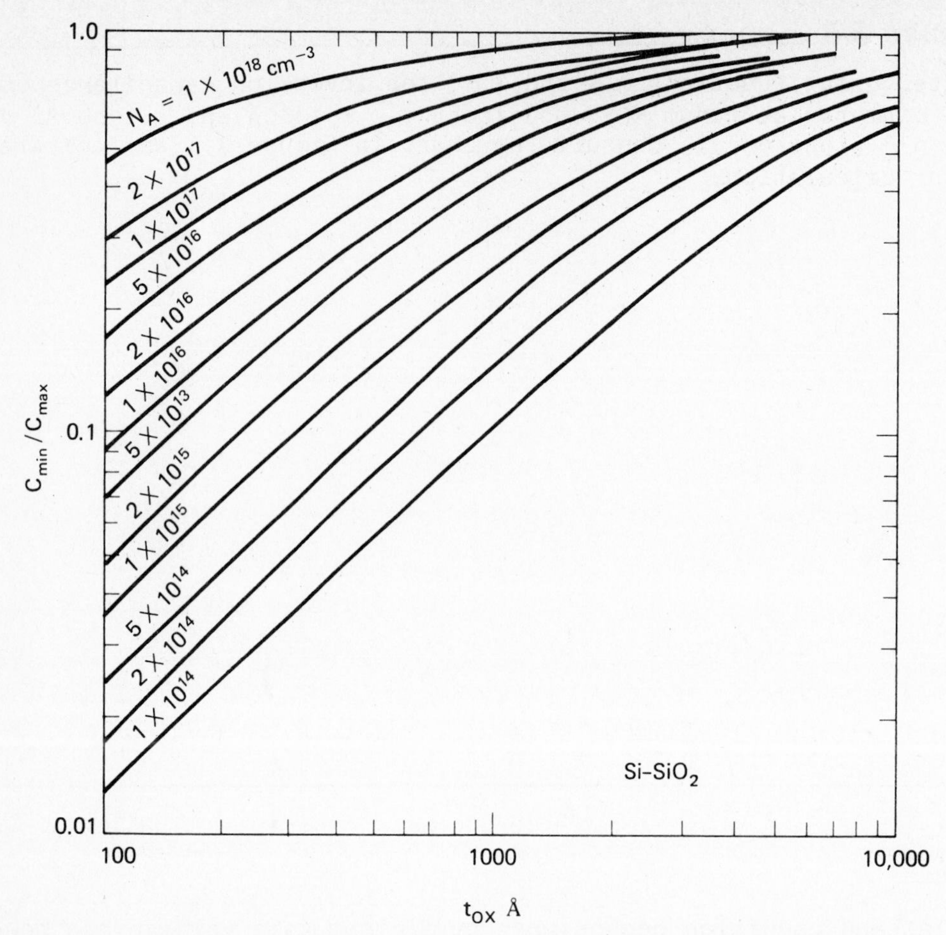

Figure 3.6 - Minimum capacitance (normalized to maximum capacitance) versus oxide thickness for an MOS capacitor for various doping levels. At high frequencies this minimum capacitance is also the inversion capacitance.

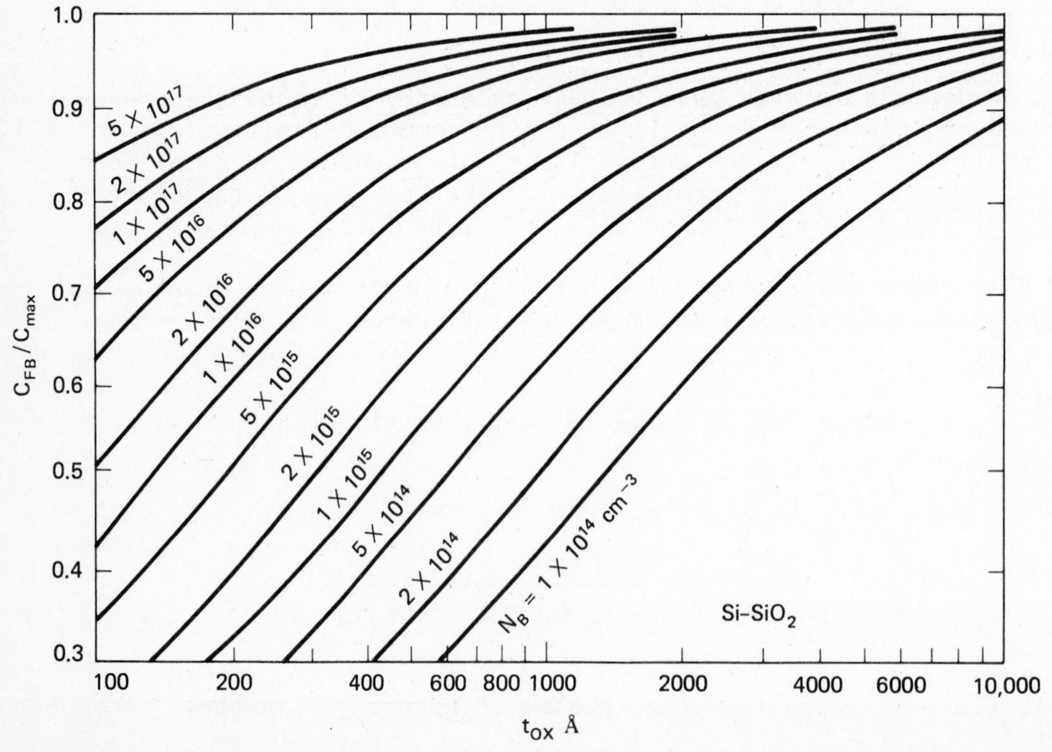

Figure 3.7 - Normalized flatband capacitance versus oxide thickness for various doping levels for an ideal MOS capacitor.

Exercise 3.4

Using the parameters of Exercise 3.3, compute C_{FB} for the MOS capacitor and compare to that given in Figure 3.7.

The flatband voltage is easily recognized and measured by observing the shift in the C-V curve at the flatband capacitance level (see insert in Figure 3.9).

Exercise 3.5

Consider the C-V curve shown in Figure 3.8. The device is a MOS capacitor. Answer the following:

a) Is the silicon n or p type? Why?

b) Identify the "low frequency" and "high frequency" portions of the curves. Give reasons for your choices.

c) Identify inversion, depletion and accumulation bias ranges. Give reasons for your choices.

d) What is C_{min}/C_{max} ratio? If the oxide thickness is measured to be 1000 Å, find the doping level N_B. What is the resistivity? (Use the appropriate chart from Case 1.)

e) Find C_{FB}/C_{max} from the chart in Figure 3.7. Use this value and find V_{FB} from the C-V curve.

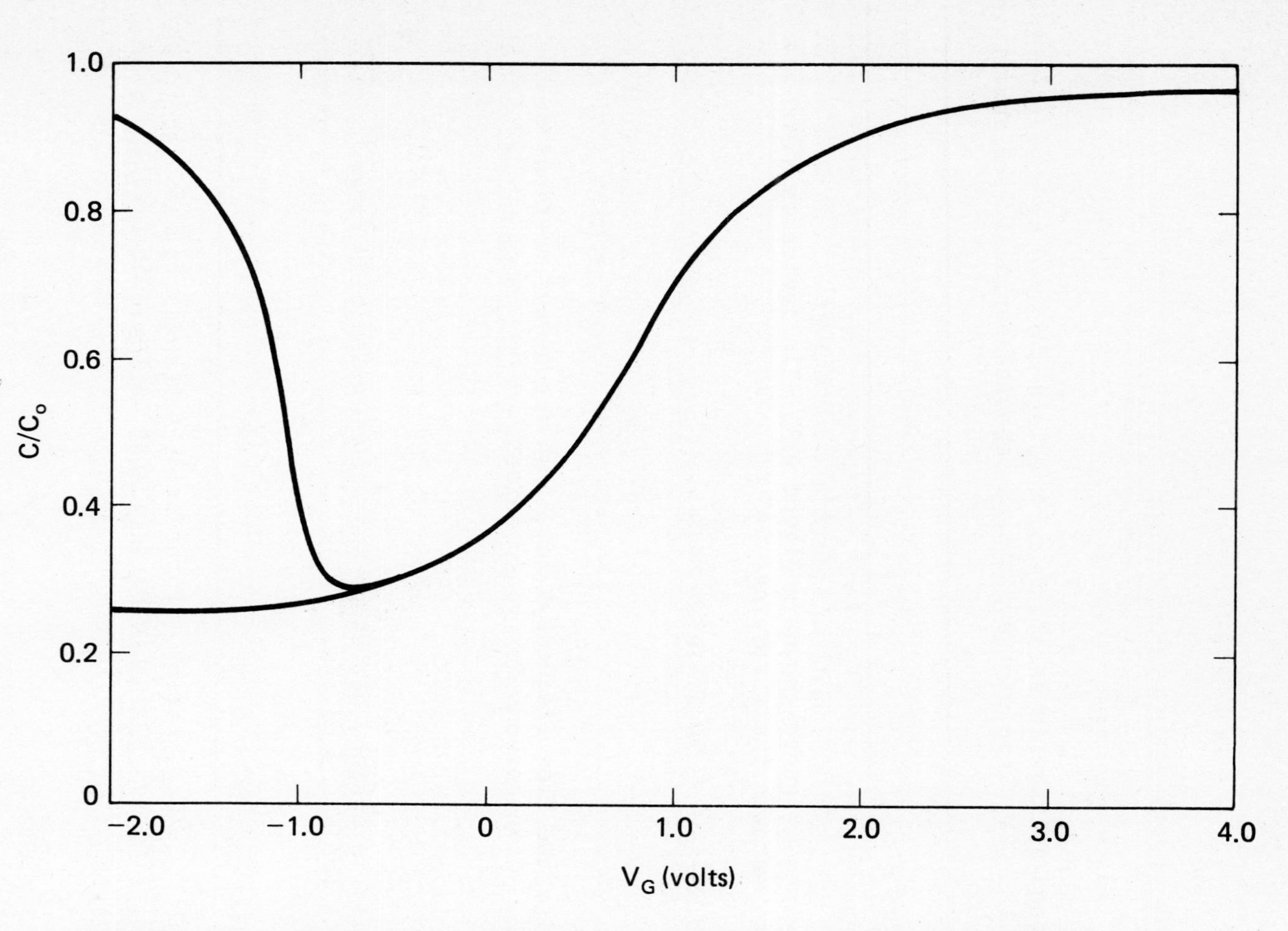

<u>Figure 3.8</u> - MOS C–V curve showing high and low frequency measurements. See Exercise 3.5.

In going through this exercise we see that there is a certain method to the analysis of a C-V curve. This procedure is summarized below.

1. From the measured CV curve, determine whether it is p or n type silicon.

2. Find accumulation and obtain C_{max}, find t_{ox} from the formula of Equation 3.2 if the area is given or vice versa.

3. Find inversion and obtain C_{min}, find N_B from C_{min} and t_{ox} by looking it up on the chart of Figure 3.6 or by the formulas of Equations 3.5 and 3.6.

   NOTE: If the formulas are used, then a transcendental equation for N_B is obtained, which must be solved iteratively:

$$N_B = [4KT/q^2]\ [K_s\epsilon_o/A^2]\ [C_{max}/C_{min})-1]^2 \ln N_B/n_i.$$

4. Find C_{FB} for the t_{ox} and N_B found by using the chart in Figure 3.7 or using the C_{FB} formula given in Equations 3.8 - 3.10.

5. Chart the $[C_{FB}, V_G]$ coordinate on the C-V curve given and on the $V_G = 0$ axis. The difference between $V_G = 0$ and $V_G = V_{FB}$ is the flatband voltage. Find V_{FB}.

6. Using N_B found above find ϕ_{ms} from Equation 3.1 or the chart in Figure 3.4 for Al-Si.

7. Use the formula for V_{FB} (Equation 3.7) and solve for and find Q_{SS} (and Q_{SS}/q as it is normally quoted).

Example 3.1

Consider another MOS capacitor. The measured CV curve is plotted in Figure 3.9. The folowing observations may be made using the procedure just outlined.

1. The silicon is p type.

2. Observe that $C_{max} = K_{ox}\epsilon_o A/t_{ox}$, find t_{ox}:

   $100\text{x}10^{-12}$ [F] = 3.9 x $8.86\text{x}10^{-14}$ x $2.89\text{x}10^{-3}$ / t_{ox}.

Answer: t_{ox} = 998.6 Å.

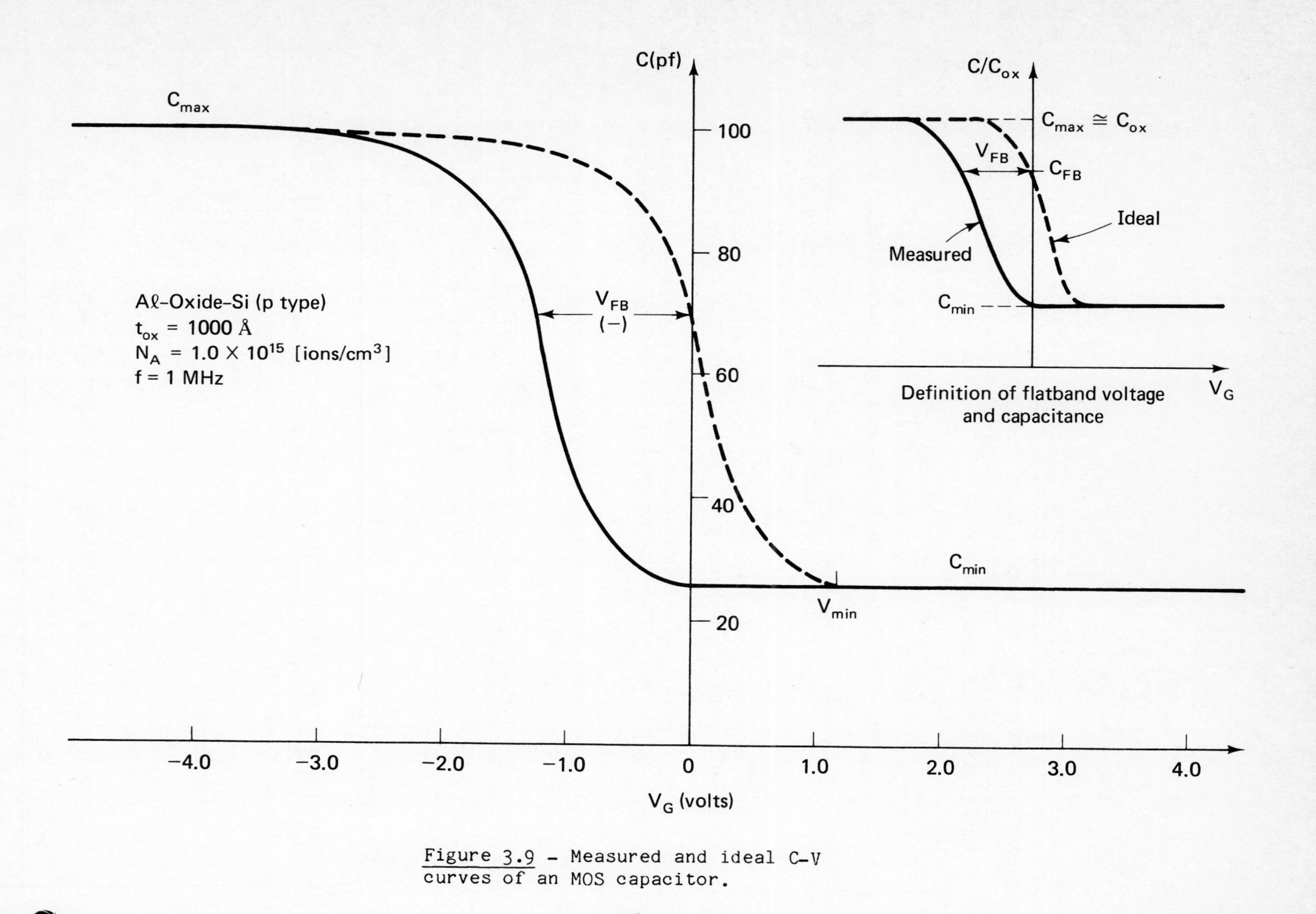

Figure 3.9 - Measured and ideal C–V curves of an MOS capacitor.

3. Find N_B from the given curves; $t_{ox} = 1000Å$, $C_{min} = 25$ [pF], $C_{min}/C_{max} = 0.25$.

Answer: $N_B = 10^{15}$ [cm^{-3}].

4. Find C_B/C_{max} from the given curves. Observe that $N_B = 10^{15}$ [cm^{-3}], $t_{ox} = 1000Å$; $C_{FB}/C_{max} = 0.7$.

Answer $C_{FB} = 70$ [pF].

5. Find V_{FB} from the measured data (Figure 3.9).

Answer $V_{FB} = -1.25$ [V].

6. Find ϕ_{ms} from charts; n type Si, $N_B = 10^{15}$ [cm^{-3}].

Answer: $\phi_{ms} = -0.85$ [volts].

7. Find Q_{SS}; use $V_{FB} = \phi_{ms} - [Q_{ss}/(Cox/A)]$

$$-1.25 \text{ [v]} = -0.85 \text{ [v]} - [Q_{ss} / (100 \text{ [pF]}/2.83\times10^{-3}[cm^2])];$$

Answer: $Q_{ss} = 12.8\times10^{-9}$ [$coul/cm^2$] and

$Q_{ss}/q = 8\times10^{10}$ [$charges/cm^2$].

Exercise 3.6

Consider the measured C-V curve shown in Figure 3.10. The available data on the device is:

a) Al gate;

b) SiO_2 insulator, $t_{ox} = 988Å$ (measured ellipsometrically);

c) Si resistivity, = 10-15 [ohm cm] (mfg. spec.), boron doped;

d) Area = 120 [μm] x 120 [μm];

e) Measurement frequency = 1 [MHz].

Answer the following:

a. Is the silicon n or p type? How does it compare with the impurity type quoted?

b. From the given area and measured C_{max}, how does the oxide thickness compare to that measured? What is the percent difference between the given t_{ox} and your calculations?

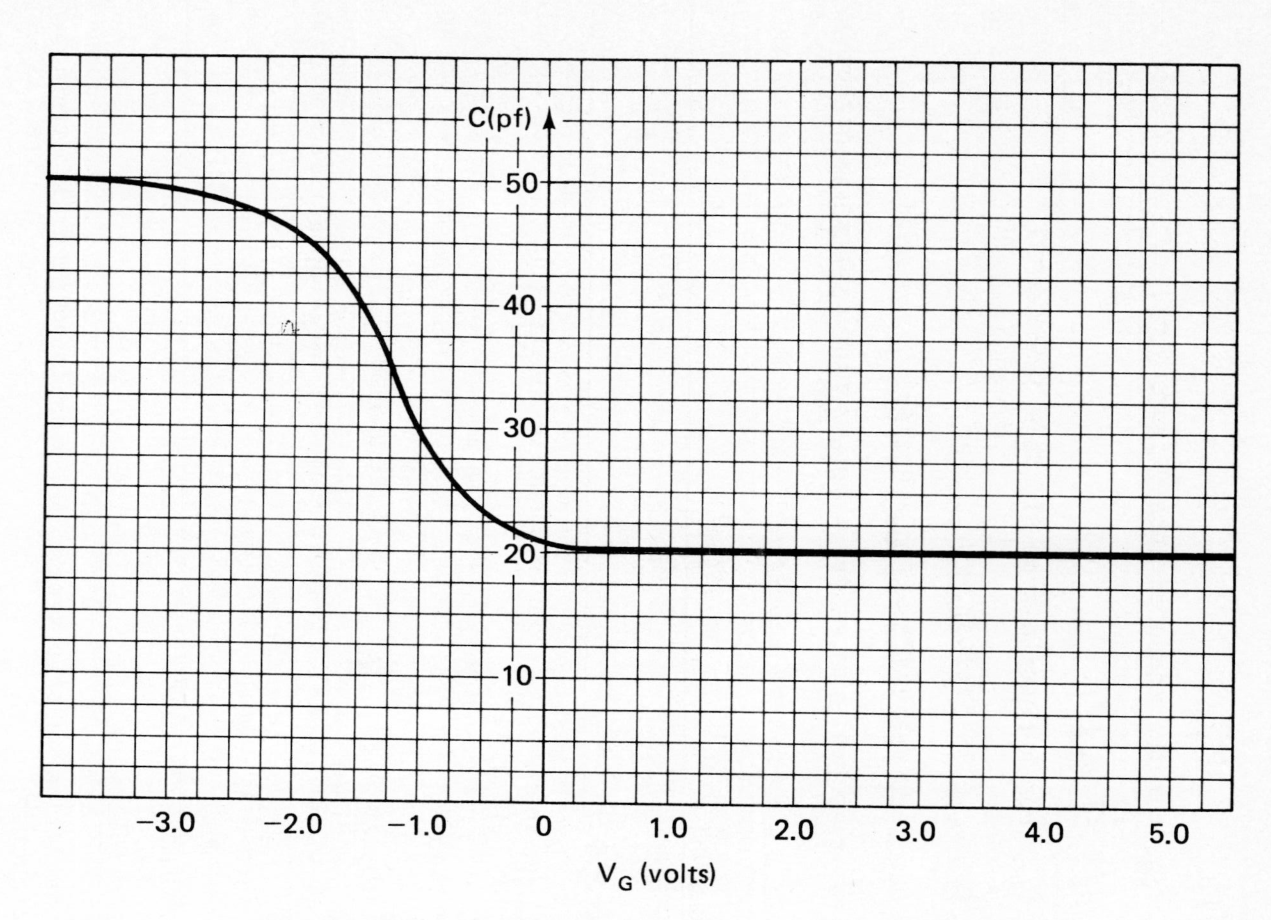

Figure 3.10 - MOS C–V curve for Exercise 3.6.

c. Assuming the impurity distribution to be uniform, calculate what the doping concentration should be. What resistivity does this correspond to? How does it compare to that quoted by the manufacturer?

d. What is the flatband capacitance?

e. What is the flatband voltage?

f. What is the aluminum-silicon work function difference?

g. What is the effective interface charge density, Q_{SS}?

Computer Laboratory

Write a computer code to implement the analysis sequence given above. Validate the program with the example given and with the data of Exercise 3.6.

OBSERVATIONS

From the charts of C_{min} versus doping and oxide thickness and from the equation thus far given we may observe that:

1. C_{max} increases as t_{ox} decreases;

2. C_{min} increases as t_{ox} decreases;

3. C_{min} increases as N_B increases;

4. Larger Q_{SS} causes larger flatband voltage shifts.

Exercise 3.7

If a measured C-V curve varies 5 percent as it goes from C_{max} to C_{min} (in high frequency), for a bias swing of +/- 20 [V], what do you suspect the range of doping of the silicon to be? Why? It is given that t_{ox} = 400Å.

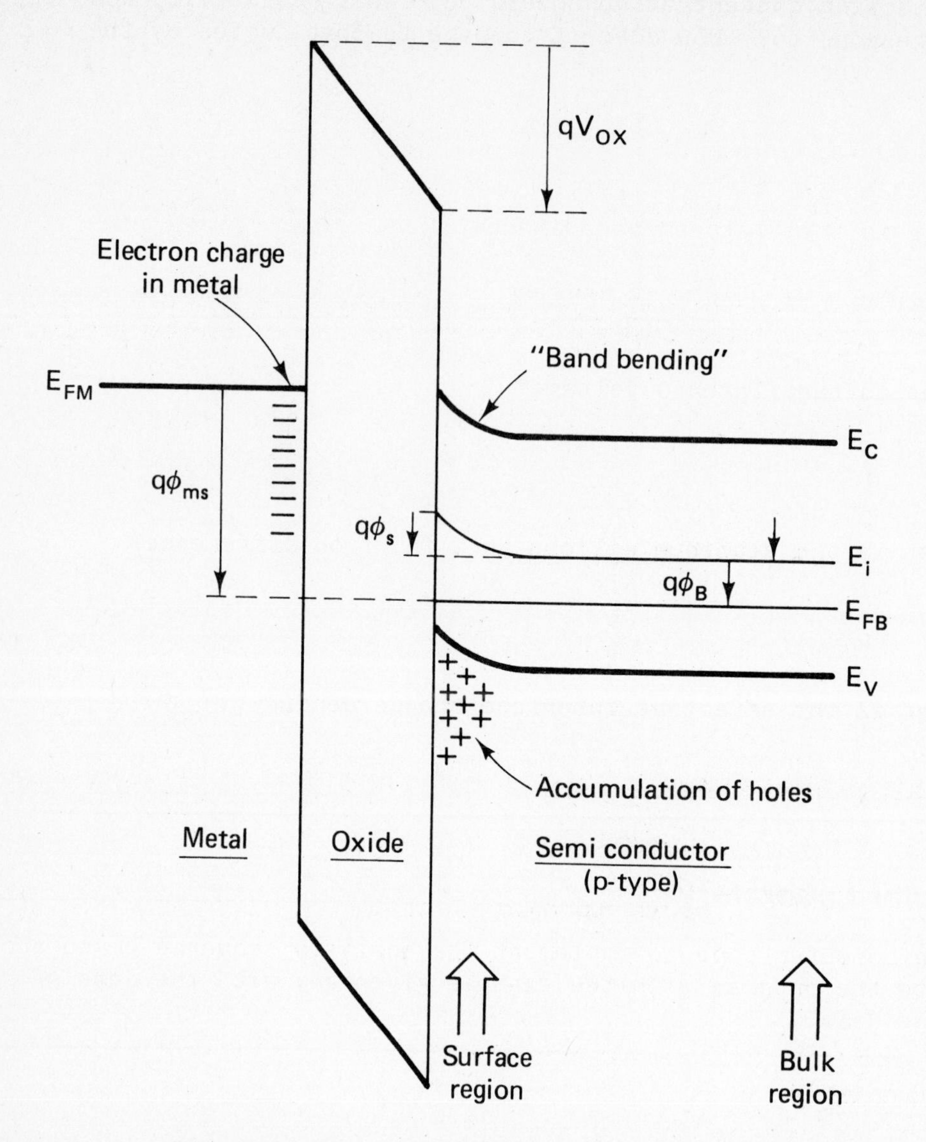

Figure 3.11 - MOS capacitor structure in accumulation showing bandbending at the surface due to difference in work functions for metal and semiconductor with no bias applied.

CASE 4: CAPACITANCE

Objectives

1. Understand how MOS C-V data may be used to obtain a doping profile.
2. Generate a computer code that implements a profiling algorithm.
3. Introduce the student to nonuniform profiles, especially ion implanted doping profiles.
4. Explore and understand the limits of a profiling technique.

Laboratory

Develop a profiling algorithm using C-V data.

References

1. W. VanGelden and E.H. Nicollian, "Silicon Impurity Distribution as Revealed by Pulsed MOS CV Measurements", J. Electrochemical Society, January 1971, pp. 138 - 141.
2. Glaser and Subak-Sharpe - Chapter 3.31, Chapter 5.5 and Chapter 11.63.

One may observe that the capacitance of the MOS capacitor in depletion is proportional to the square root of the inverse of the doping. An equation may be derived that expresses the value of doping at any value of capacitance-voltage:

$$N(W) = [2/qK_s \epsilon_o] \times 1/[dC^{-2}/dV_G] \ [cm^3] \qquad (4.1)$$

where C is in F/cm^{-2} per unit area and

$$W = K_s \epsilon_o / C_s = K_s \epsilon_o \ [1/C_m - 1/C_{ox}] \ [cm] \qquad (4.2)$$

Here C_m is the capacitance per unit area at the gate voltage V_G and W is the depletion width at that gate voltage.

Exercise 4.1

A MOS capacitor (C_{ox}/A = 21 [nF/cm^2]) biased in deep depletion yields the following data:

@ V_{G1} = -4.0 [V], C_{M1} = 3.296 [nF/cm^2];

@ V_{G2} = -4.5 [V], C_{M2} = 3.04 [nF/cm^2].

Find the average doping level and the depletion width.

For uniform profiles all that is required is to know two reasonably close C–V points, so that the derivative dC/dV is approximated properly. From this we may find N_B as in Exercise 4.1. To profile nonuniform profiles it is necesary that the formula be applied as the semiconductor is being depleted. The limitation here is that if inversion occurs before we lose our opportunity to deplete the whole profile being studied we can obtain no more information. One way to get around this is to pulse the gate voltage very quickly from accumulation into depletion. Since the inversion carriers cannot respond to a fast signal then the silicon will be in depletion while the pulse is on. Pulsing beyond the point where inversion normally occurs causes the condition known as deep depletion. Care must be exercised to keep the pulse width short. There is always some small amount of leakage, and if one holds the device biased at deep depletion long enough, an inversion layer slowly forms. Devices which have a leakage problem are not suitable for this measurement. Also devices that leak through the oxide are unsuitable and must be tested for this condition and screened out.

Pulsing is normally done from a point in accumulation, and one returns the bias there increasing the pulse voltage consecutively to sweep out a complete curve, or profile. The deep depletion characteristic is shown in Figure 4.1 and compared to the high frequency curve. It must be noted that the deep depletion condition is a transient one.

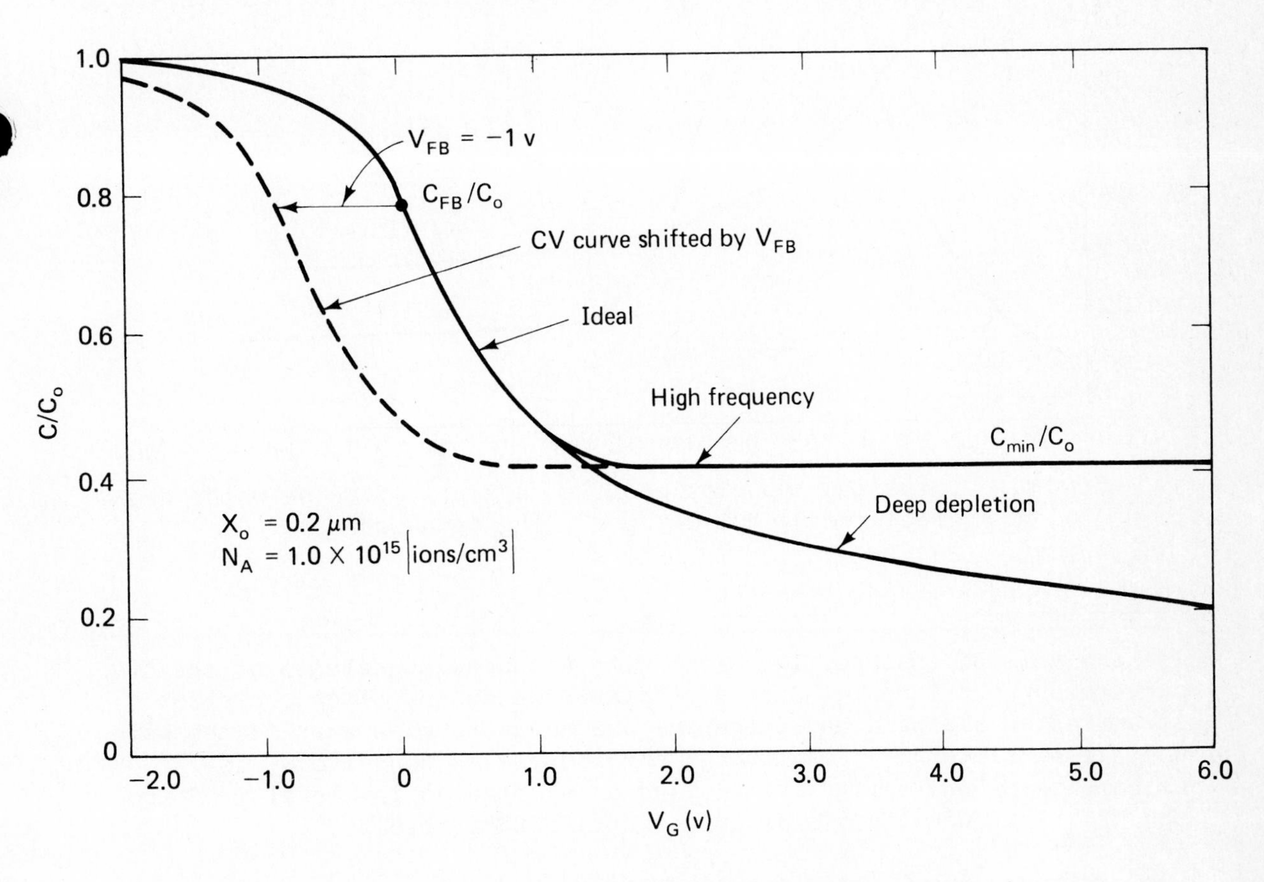

Figure 4.1 – MOS C–V curve showing deep depletion characteristic and the definition of flatband.

TYPICAL NONUNIFORM PROFILES

Profiles that have a gaussian form are typical of the form profiled by this method. Some result from the use of ion implantation in various processes (see Case 10). They have the form:

$$N(x) = N_{po} \exp - [(x-\mu)^2/2\sigma^2] + N_B \ [cm^{-3}] \qquad (4.3)$$

Two such profiles are shown in Figure 4.2 with appropriate parameters.

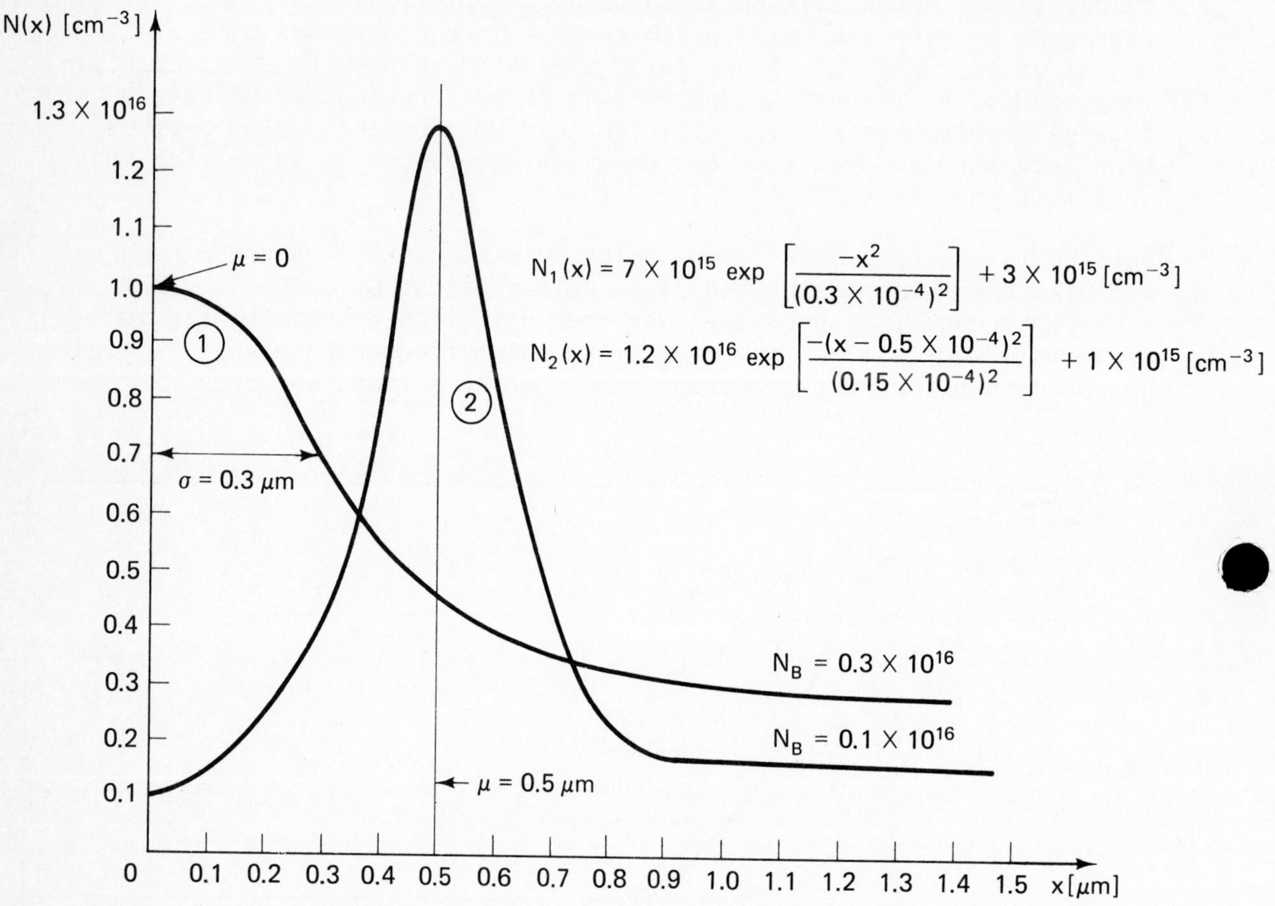

Figure 4.2 - Gaussian impurity profiles typical of those which are measured by MOS C-V profiling techniques.

ANALYSIS OF NONUNIFORM PROFILES

The analysis of the profile begins with a thorough analysis of the C-V curve along the lines of Case 3. This is necessary since depletion occurs after flatband and therefore one needs to know what is the gate voltage at flatband. Even though the profile is nonuniform, an assumption of uniformity is made and an average doping level is found for N_B. This usually yields quite satisfactory results.

The C-V data then is given as capacitance and voltages starting at flatband. The equations are modified as follows:

$$W(i) = K_s \epsilon_o \ [1/C_m(i) - 1/C_{max}] \qquad (4.4)$$

$$N(i) = \frac{2}{qK_s\epsilon_o} \left\{ \frac{1}{\dfrac{[C_m(i+1)]^{-2} - [C_m(i-1)]^{-2}}{V_G(i+1) - V_G(i-1)}} \right\} \quad (4.5)$$

This last equation is appropriate for i = 2 to the last point; for i = 1, we must use a different equation:

$$N(1) = \frac{2}{qK_s\epsilon_o} \left\{ \frac{1}{\dfrac{[C_m(2)]^{-2} - [C_m(1)]^{-2}}{V_G(2) - V_G(1)}} \right\} \quad (4.6)$$

Exercise 4.2

Analyze the C-V curve given in Figure 4.3.

a. Find t_{ox}, C_{max}, N_B, C_{FB}, and V_{FB}.

b. Starting at V_{FB}, discretize the deep depletion C-V curve in equal voltage intervals of 0.5 [V]. Generate a table of capacitance versus voltage.

c. Use Equations 4.4 -4.6 to generate a profile versus depth table. Plot it and add it to this workbook.

Computer Laboratory

Develop and code an algorithm that will have a discretized C-V curve as its input and will automatically generate the profile versus depth. Make sure the program outputs a table that contains C, W, and N as a function of gate voltage. Analyze the data of Exercise 4.2 to validate your algorithm. Analyze and enter into this workbook the analysis of the curve given in Figure 4.4. Include a copy of the plot of the profile.

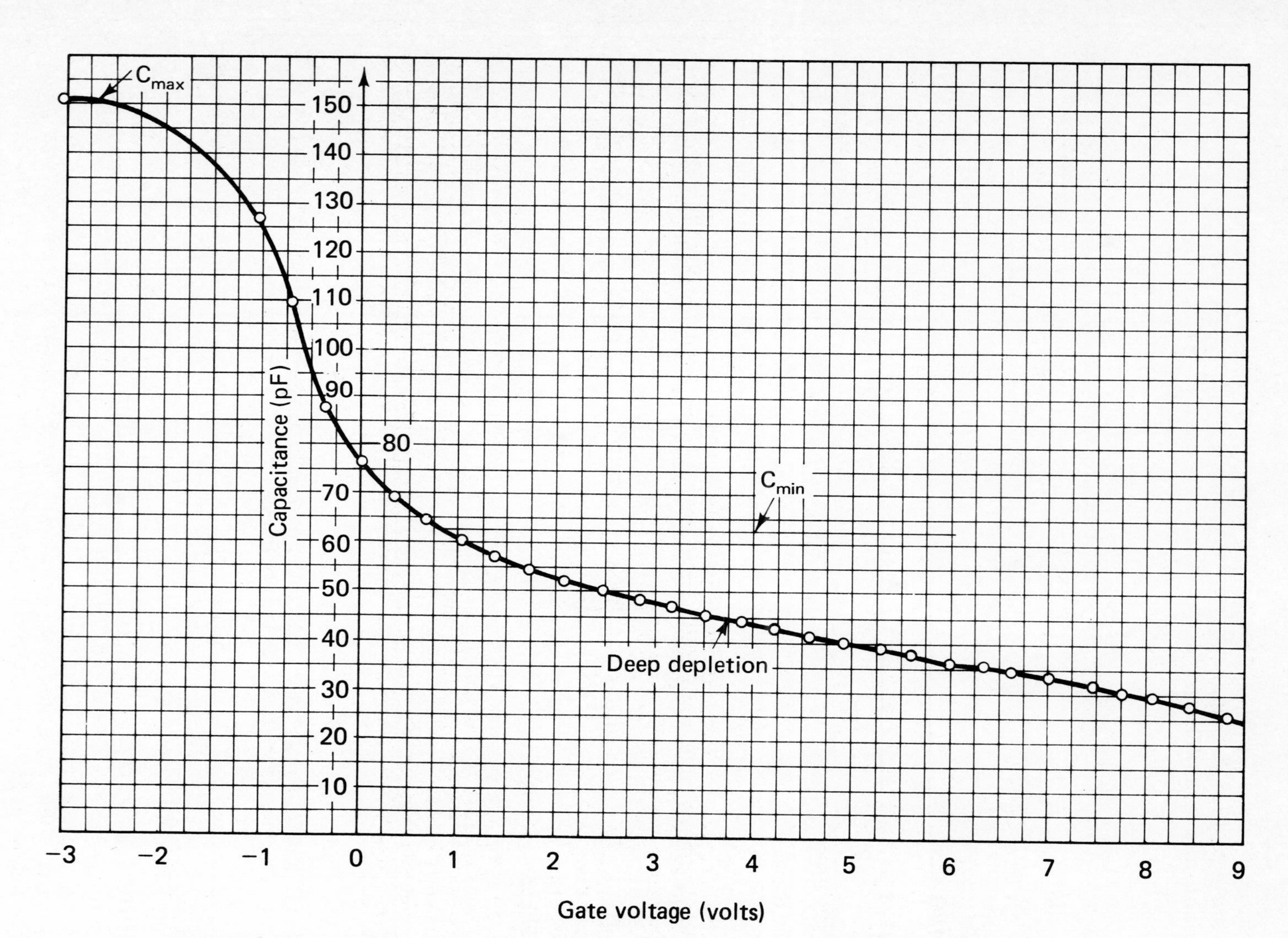

Figure 4.3 – MOS deep depletion capacitance–voltage data used in Exercise 4.2. The device area is $A = 1.63 \times 10^{-3}$ [cm^3], C_{max} = 149.85 [pF] and C_{min} = 62.67 [pF].

C_{ox} = 112.0 [pF]
C_{min} = 16.0 [pF]
V_{FB} = 1.3 [volts]
Area = 5×10^{-4} [cm^{-2}]
t_{ox} = 460Å

V_G [volts]	C_m [pF]
-1.3	32.0
-1.1	30.0
-0.9	28.2
-0.7	26.4
-0.5	24.8
-0.3	23.4
-0.1	22.2
0.1	21.1
0.3	20.2
0.5	19.3
0.7	18.5
0.9	17.8
1.1	17.2
1.3	16.5
1.5	15.9
1.7	15.4
1.9	14.8
2.1	14.4
2.3	13.8
2.5	13.4
2.7	12.9
2.9	12.4
3.1	11.9
3.3	11.5
3.5	11.1
3.7	10.6
3.9	10.1
4.1	9.6
4.3	9.0
4.5	8.4
4.7	7.7
4.9	6.9
5.1	6.3
5.3	5.4
5.7	4.8
5.9	4.4
6.1	4.0
6.3	3.8
6.5	3.5
6.7	3.3
6.9	3.2

Figure 4.4 - MOS deep depletion capacitance-voltage data used in the computer laboratory exercise.

OBSERVATIONS

Errors are easily introduced when the capacitance does not vary very much as a function of voltage. Thus a practical upper level doping limit is 10^{18} [ions/cm^2]. Most semiconductors in the IC business are doped above 10^{14} [ions/cm^2], although there are devices that make use of very lightly doped material. As far as oxide thickness is concerned, the thinner, the better for MOS capacitance profiling. If the device has already been manufactured then a practical upper limit is 5000Å. One may always analyze thicker capacitors but a greater error is expected.

CASE 5: MOSFET

Objectives

1. Understand the components of threshold of a MOSFET.
2. Understand the effect of substrate bias, analyze and generate substrate sensitivity curves.
3. Develop a threshold model.
4. Develop a subthreshold model.

Laboratory

Develop and code algorithms to model the threshold of a MOSFET.

References

1. Richman, Chapter 2.
2. Grove, Chapter 11.
3. Glaser and Subak-Sharpe, Chapter 3.4.
4. V.L. Rideout, F.H. Gansslen and A. Leblanc, "Device Design Considerations for Ion Implanted n-Channel MOSFETS", IBM Journal of Research and Development, Jan. 1975, pp. 50 -59.

THRESHOLD VOLTAGE

Consider the MOSFET cross-section drawn below. For conduction between source and drain to occur when a source to drain voltage is applied, there must be an inversion region created that will connect the two. Basically the source and drain doped regions are reverse biased p-n junctions and only a leakage current will flow through them below threshold. If the inversion region is created at the surface between source and drain when a gate to substrate voltage is applied, then a current will flow and we will have control of the flow. The traditional model of threshold assumes no current flow before a certain voltage V_T (threshold voltage) is applied to the gate and a controlled amount of current flows when at least this voltage is applied. The MOSFET with threshold voltage applied and channel created is shown in Figure 5.1.

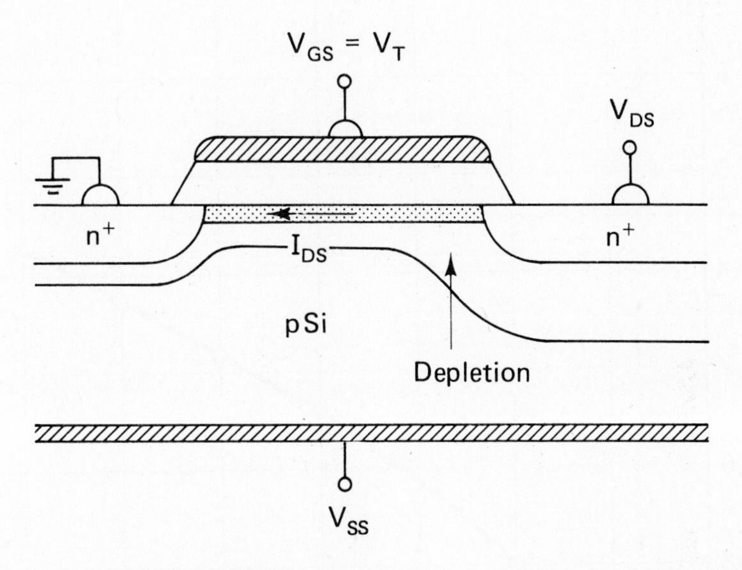

Figure 5.1 - N-Channel MOSFET biased at threshold showing the created inversion layer and drain current flowing.

Reverse biasing the substrate with respect to the source will cause it to be harder to create the inversion layer at the surface thereby increasing threshold. Work function difference and oxide charge also affect threshold in the same manner that they affected the shifting of the C-V characteristic of the MOS capacitor. Similarly, all elements of threshold may be traced to creation of an inversion layer in that MOS capacitor:

$$V_T = V_{FB} + 2\phi_B + \sqrt{\frac{2 K_S \epsilon_o N_B q (2\phi_B + V_{SS})}{C_{ox}/A}} \qquad (5.1)$$

This equation has been derived for n-channel enhancement MOSFETs.

The variation of threshold V_T with substrate bias V_{SS} when plotted yields a very important characteristic: the substrate-sensitivity curve for the MOSFET. An example of such a characteristic is plotted in Figure 5.2.

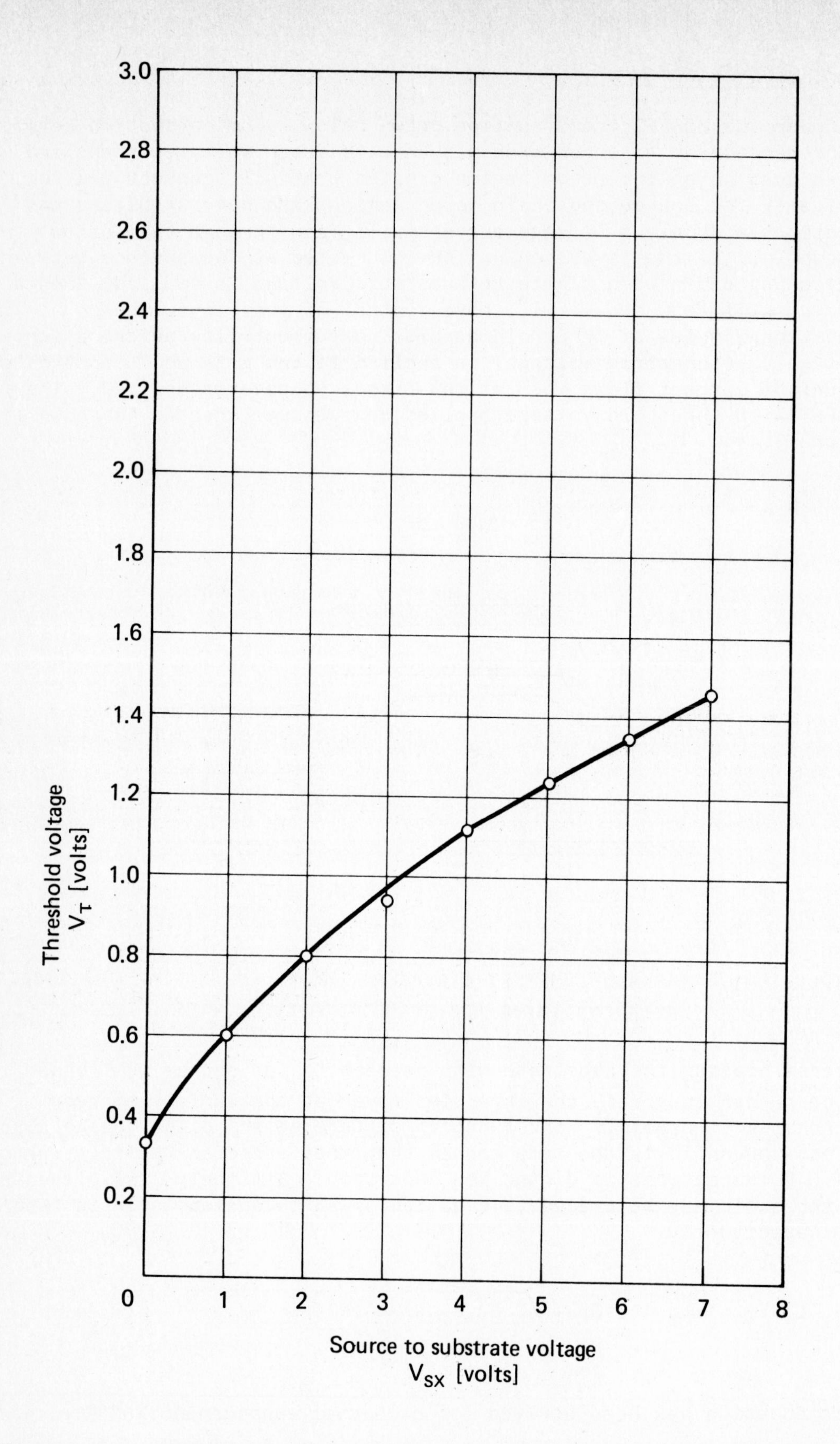

Figure 5.2 - Substrate sensitivity curve of an n-channel MOSFET manufactured on 2 [ohm cm] p Si with a 430Å SiO_2 gate insulator and of 20 [μm] channel length.

Exercise 5.1

Using Equation 5.1 compute the substrate sensitivity curves for the following four cases (assume n channel MOSFETs with Al gates and Q_{SS}=0.0):

Case	1	2	3	4
Substrate doping	12 [ohm cm]	12 [ohm cm]	2 [ohm cm]	2 [ohm cm]
Oxide thickness	500Å	1000Å	500Å	1000Å

	CASE			
	1	2	3	4
V_{SS} [volts]	V_T [volts]	V_T [volts]	V_T [volts]	V_T [volts]
0.0				
1.0				
2.0				
3.0				
4.0				
5.0				
6.0				
7.0				
8.0				
9.0				
10.0				

Plot these four curves on one sheet of graph paper and add it to this workbook.

This characteristic is usually supplied by device or process designers to circuit designers so they may design circuits to be insensitive to threshold variations due to substrate to source bias changes. Since circuits often are operated at fixed substrate to ground bias, one may easily find the operating threshold voltage for this device.

One may analyze measured threshold substrate sensitivity curves for flatband and Q_{SS} by using Equations 5.1 and 3.7 and the following procedure:

1. Choose a particular substrate bias;
2. Read the corresponding threshold from the data;
3. Use Equation 5.1 with the available device data to find V_{FB};
4. Use Equation 3.7 to find Q_{SS}.

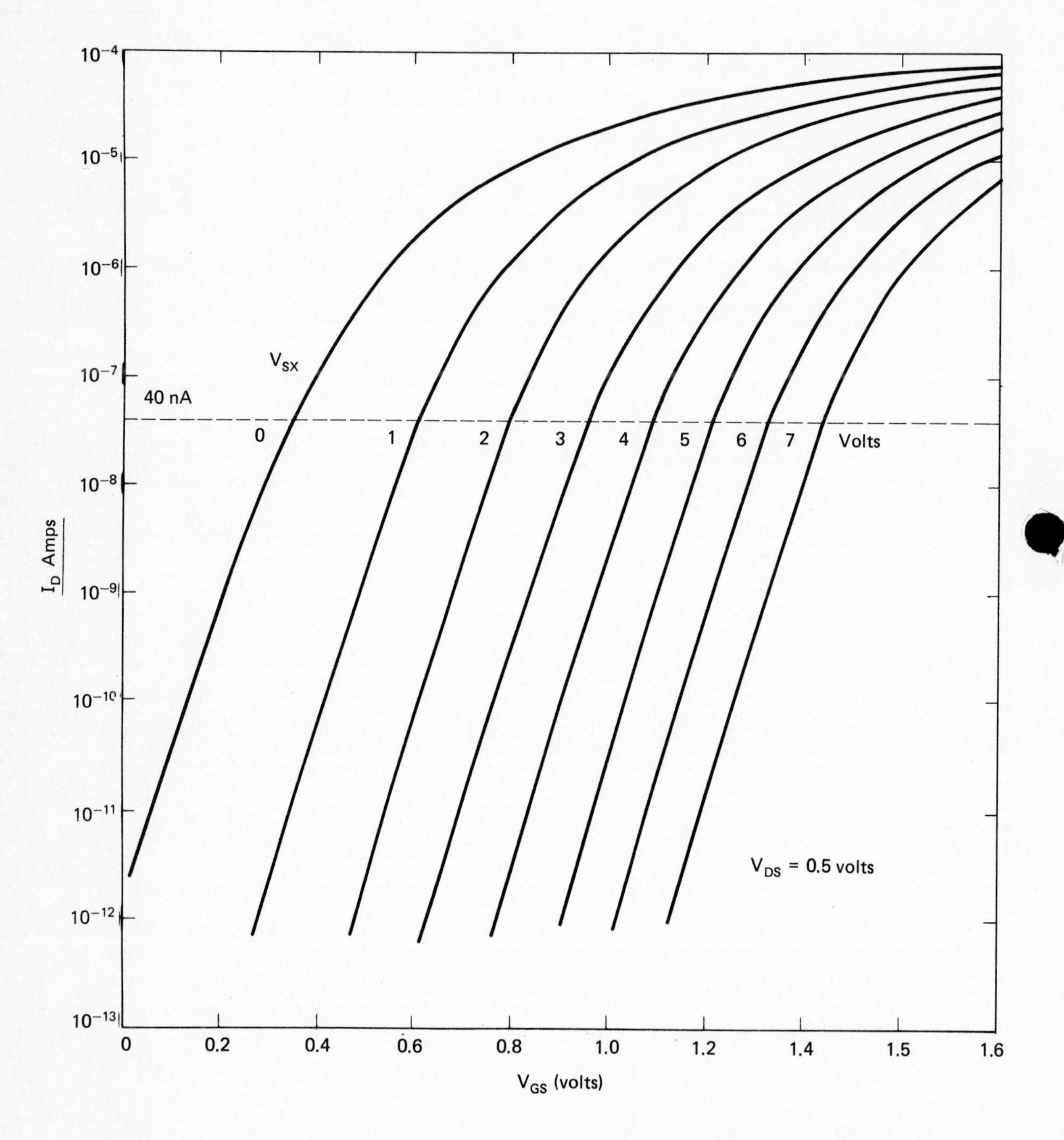

Figure 5.3 - Subthreshold current versus gate bias and definition of threshold at a certain level of subthreshold current (40 [nA] W/L).

Exercise 5.2

Consider the measured substrate sensitivity data plotted in Figure 5.2. Choose a source to substrate bias and, following the above procedure, find the threshold voltage, flatband voltage and Q_{SS} (the device data is given in Figure 5.2).

Using the given device data, compute and plot on the same graph as Figure 5.2 the theoretical substrate sensitivity curve for this device for zero flatband voltage.

MEASURING THE THRESHOLD VOLTAGE OF THE MOSFET

There are several ways of measuring threshold. One approach is from the on state of the device using the I_D - V_{DS} saturation region characteristics. This is useful for circuit design purposes and is discussed in Case 11. In device characterization work another approach is to measure drain to source current below threshold, called subthreshold current, as a function of gate bias. Threshold in this case is defined as a certain level of current flow corresponding theoretically to onset of inversion, I_{DS} = 40 [nA] x Width/Length.

Example 5.1

Measured drain current versus gate voltage characteristic for the device are given in Figure 5.3. This device has a width to length ratio of 1:1, thus drawing a line at I_{DS} = 40 [nA] we find V_T at each value of substrate bias. This generated the substrate sensitivity curve of Figure 5.2 which has been studied before.

As is observed, subthreshold current is exponentially related to gate bias below threshold. Greater understanding of threshold may be obtained by consulting Reference 5.4.

Exercise 5.3

The threshold of a certain metal gate n-channel MOSFET is characterized by the subthreshold measurements shown in Figure 5.4. The gate insulator is 430Å of SiO_2; the channel length is 19.1 [μm] and the channel width is 28.9 [μm]. Find the substrate sensitivity characteristic from this measurement. Plot it superimposed on the graph produced in answer to Exercise 5.1. Compare the slope of this curve to those already plotted. Estimate the substrate doping and resistivity.

With this doping, use Equation 5.1 and estimate the flatband voltage and Q_{SS} (V_{SS} = 0.0 [V]).

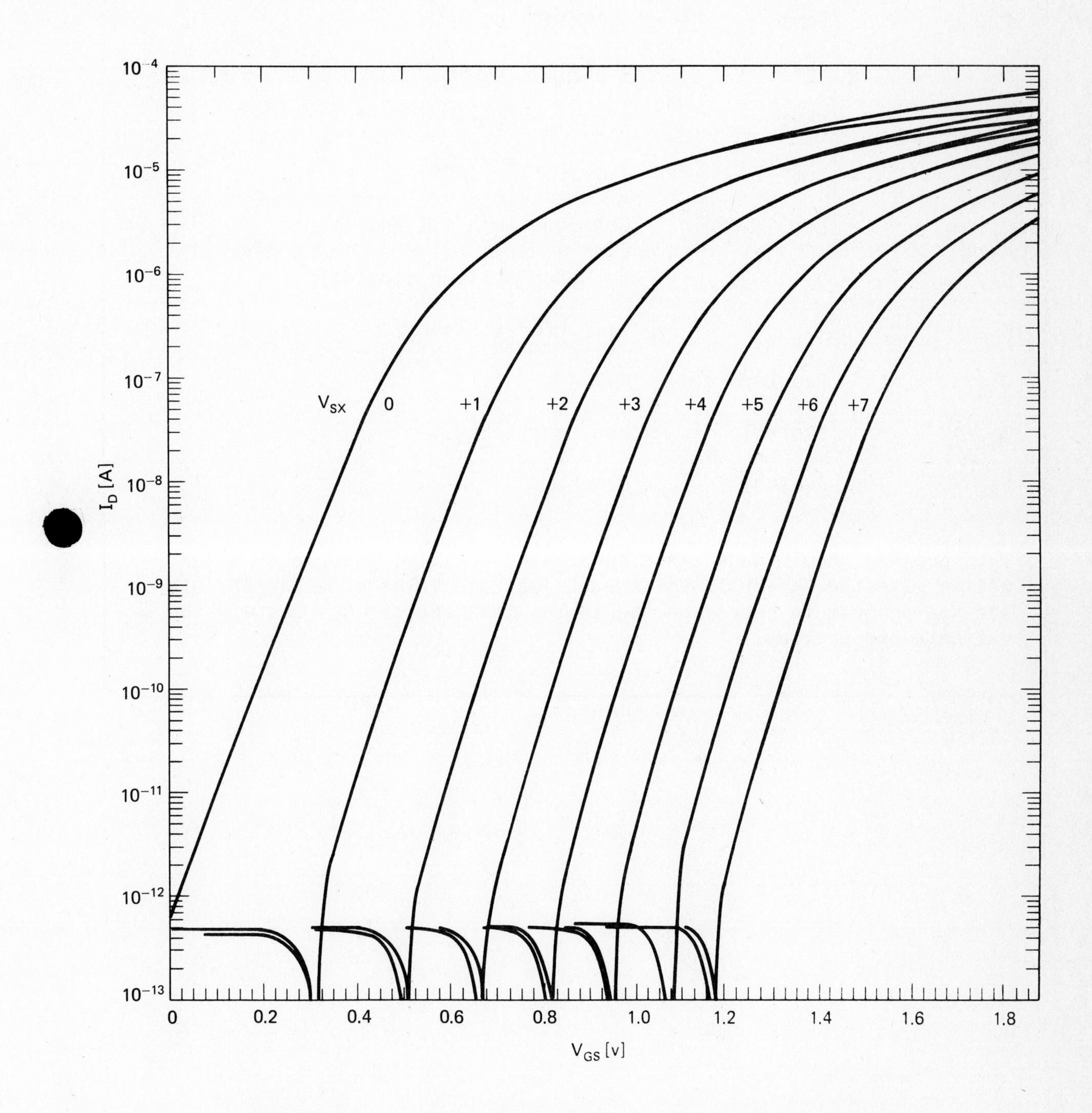

Figure 5.4 - Subthreshold current data for Exercise 5.3.

Computer Laboratory

Write a computer program encoding Equation 5.1. The following parameters of the equation should have a default value (as given) or be changed at the user's option:

Parameter	Units	Default
t_{ox}	[cm]	0.05×10^{-4} [cm]
W	[cm]	1.00×10^{-2} [cm]
L	[cm]	1.00×10^{-2} [cm]
N_B	[cm^{-3}]	$1.00 \times 10^{+-5}$ [cm^{-3}]
μ_p, μ_n	[cm^2/V sec]	Use Equations 1.5 and 1.6
K_{ox}	-	3.9 (for silicon dioxide)
K_s	-	11.7 (for silicon)
Q_{SS}	[coul/cm^2]	0.0
ϕ_{ms}	[volts]	Use Equation 3.1
V_{SS}	[volts]	0.0

This program should generate a substrate sensitivity curve or as an option give the threshold voltage at just one value of substrate bias. Use the program to reproduce the tables in Exercises 5.1 and 5.2 to validate the program.

CASE 6: DIODE

Objectives

1. Understand the mechanisms of current conduction across reversed biased and forward biased p-n junctions.

2. Understand the sources of leakage across a reverse biased junction.

3. Develop and use a model of current flow across a diode suitable for computer aided analysis.

4. Understand junction breakdown.

Laboratory

Develop and code an algorithm that models current flow in a p-n junction diode.

References

1. Grove, Chapter 6.

2. Sze, Chapter 3.

P-N JUNCTIONS

Two elements of importance in analyzing MOS integrated circuits, as far as junctions are concerned, are the reverse biased current and the junction capacitance. In this case we shall examine the current-voltage characteristic of a p-n junction. We shall assume we have what is referred to as an abrupt junction, with one side highly doped with respect to the other. A typical junction in an MOS device is the source or drain diffusion, for example. We show a schematic of one such junction in Figure 6.1.

Consider a reverse bias applied to the junction. As shall be discussed more thoroughly in the next section, the depletion region widens. A small amount of current flows: the reverse bias current I_R (see Figure 6.2). There are two contributions to current flow, a diffusion current and a generation current. Diffusion arises from the concentration gradient of carriers from one side of the junction to the other. Because of the high doping on one side with respect to the other, this diffusion mechanism is dominated by one type of carrier, usually the bulk minority carrier in the lighter doped side.

In general:

$$I_{diff} = \underbrace{q\, D_n\, n_{po}\, A_j/L_n}_{\text{(in p side)}} + \underbrace{q\, D_p\, p_{no}\, A_j/L_p}_{\text{(in n side)}} \text{ [Amps]} \qquad (6.1)$$

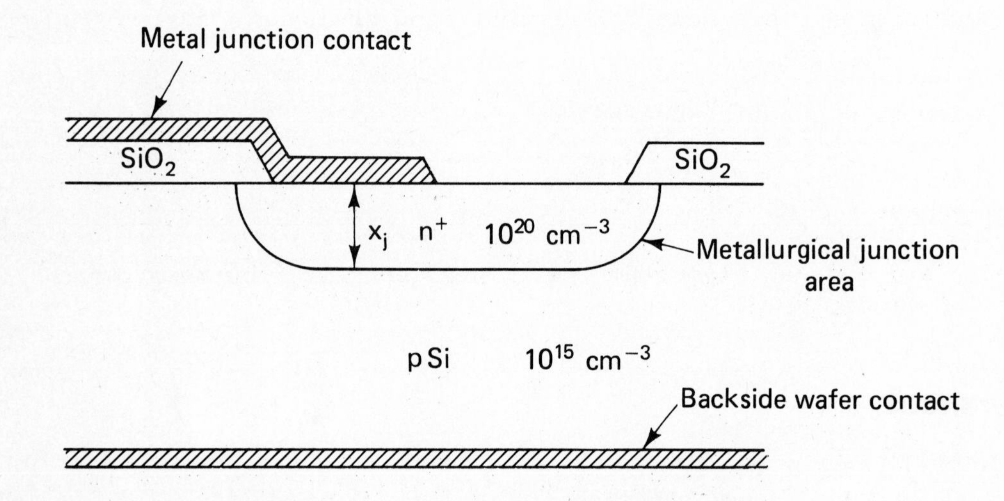

Figure 6.1 - Abrupt p-n junction fabricated with a planar process, typically a source or drain of a MOSFET is shown.

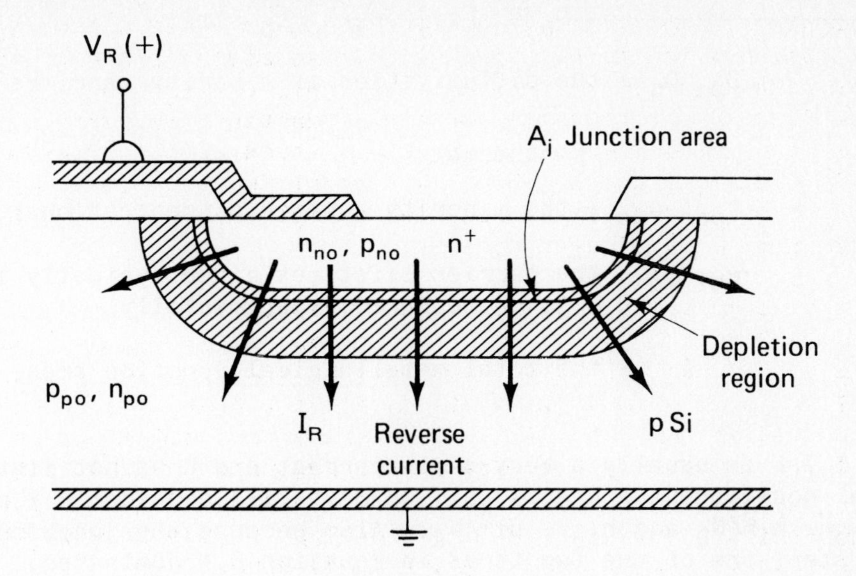

Figure 6.2 - Junction of Figure 6.1 under reverse bias showing depletion regions and reverse current flow.

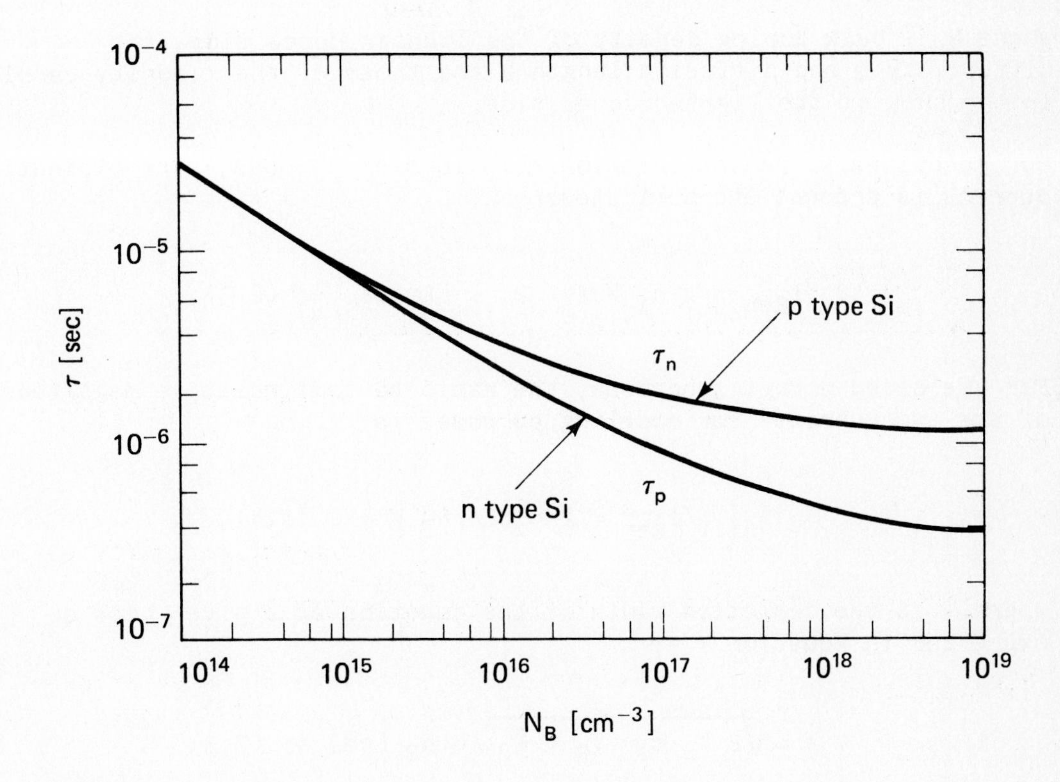

Figure 6.3 - Minority Carrier lifetime versus doping density for silicon at 300°C.

where: L_n, L_p = the minority carrier diffusion lengths

$= \sqrt{D_n \tau_n}$ [cm] or $\sqrt{D_p \tau_p}$ [cm];

and: D_p, D_n = the diffusivities of minority carriers

$= \mu V_{thermal} = \mu K T / q$ [cm^2/V sec];

n_{po}, p_{no} = the minority carrier concentrations;

τ_n, τ_p = the carrier lifetimes given typically in Figure 6.3, or measured experimentally;

A_j = the total metallurgical junction area.

Since I_{diff} is usually a very small current and does not disturb thermal equilibrium a great deal, we may still consider $n \times p = n_i^2$, and $p_{no} = n_i^2/N_D$ and $n_{po} \doteq ni^2/N_A$. Also because the junction is a one sided step, one of the two terms in Equation 6.1 dominates. This yields:

$$I_{diff} = q\, D\, n_i^2\, L\, A_j / N_B \text{ [Amps]} \quad (6.2)$$

where N_B = bulk doping density of the lighter doped side, the diffusivity D and diffusion length L are those of the minority carriers in the bulk of the lighter doped side.

For junctions where the lifetime τ is in microseconds, a recombination current is present and dominates:

$$I_{gen} = q\, n_i\, W\, A_j / 2 \quad \text{[Amps]} \quad (6.3)$$

For one sided abrupt junctions, the ratio to test relative magnitudes of the two currents for modeling purposes is:

$$I_{diff}/I_{gen} = 2\, n_i \tau L / N_B\, W \quad (6.4)$$

where W is the depletion width of the junction at a given bias as expressed in Equation 7.3:

$$W = \sqrt{2\, K_s\, \epsilon_o\, (V_R + \phi_B)/q\, N_B} \text{ [cm]} \quad (7.3)$$

Example 6.1

To calculate the total junction area of a planar junction is often an important problem. We repeat here below some of the important features as a do-it-yourself pursuit.

Consider a junction that was fabricated by diffusing the impurity into the surface of the wafer with a properly etched oxide layer as a mask, such as drawn below:

Figure 6.4 - Cutaway showing a planar diffused junction for purposes of calculating junction area.

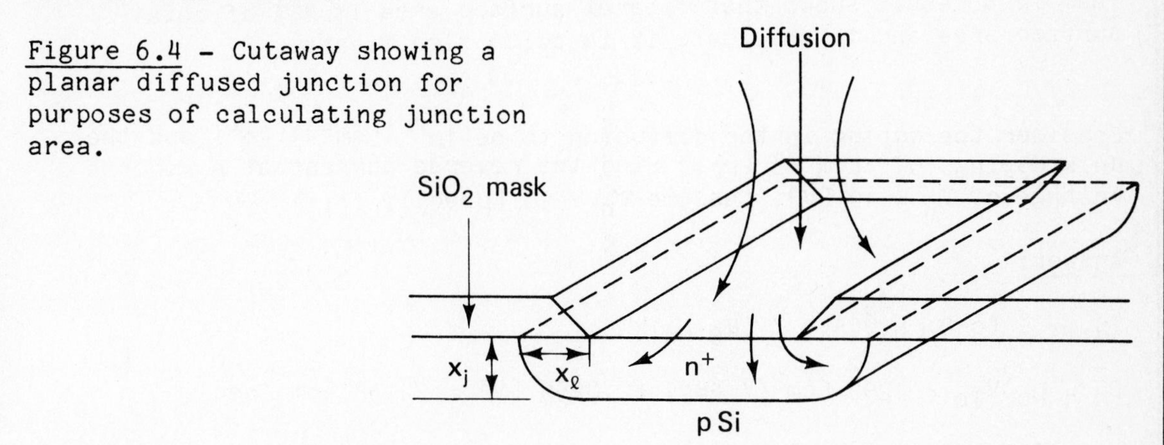

The area of the junction is computed by adding the surface areas of the components shown below:

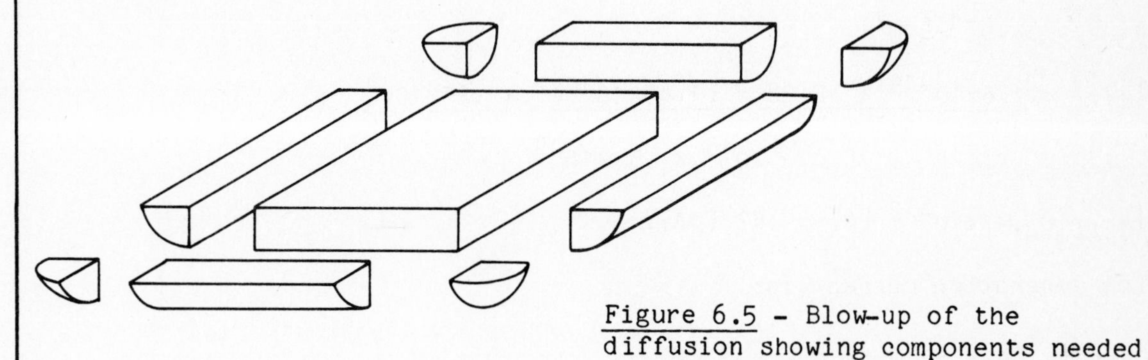

Figure 6.5 - Blow-up of the diffusion showing components needed to calculate surface area of junction A_j.

Use x_j = 1.6 [μm]; x_L = .75 x_j = 1.2 [μm]; W = 20 [μm]; L = 100 [μm];

$A_j = A_s(4x1/4 \text{ cylinders}) + A_s(4x1/8 \text{ spheres}) + A_s(\text{planar area});$

$A_s \text{ (planar)} = [W - 2(x_j - x_L)] \times [L - 2(x_j - x_L)]$

$= .19 \times 10^{-4} \text{ [cm}^2\text{]};$

$A_s \text{ (cylinders)} = 2 \times 1/4 \times 2\pi X_j \times L + 2 \times 1/4 \times 2 * X_j W$

$= \pi X_j (W + L)$

$= \pi \times 1 \text{ [μm]} \times 120 \text{ [μm]} = 6.03 \times 10^{-6} \text{ [cm}^2\text{]};$

A_s (spheres) $= 4 \times 1/8 \times 4/3\,\pi\, X_j^2 = 5.36\times10^{-8}$ [cm^2];

A_j $= .25\times10^{-4}$ [cm^2];

A_s planar $= .19\times10^{-4}$ [cm^2];

A_s/A_j $= .75$.

This last result shows that lateral surface area is 25% of total surface area and in this case it is quite significant.

Consider the doping in the diffusion to be 10^{20} [cm^{-3}] (n^+) and the bulk doping 10^{15} [cm^{-3}] (p). Find the reverse current at a reverse voltage of $V_R = 10$ [V], (assume $\tau_n = 10$ [μsec]).

Answer:

$I_{diff} = (q\, D_n\, n_i^2\, A_j) / (N_B\, L_n)$;

$D_n = \mu_n V_{Th} = 1250$ [cm^2/V sec] x .0259 [V] = 32.38 [cm^2/sec].

$L_n = \tau_n D_n$ = 10 [μsec] x 32.38 [cm^2/sec] = $.57\times10^{-4}$ [cm] = .57 [μm]

$N_B = 10^{15}$ [cm^{-3}]

$A_j = .25\times10^{-4}$ [cm^2]

$$I_{diff} = \frac{1.6\times10^{-19} \times 32.38 \times (1.45\times10^{10})^2 \times .25\times10^{-4}}{10^{15} \times .57\times10^{-4}};$$

$I_{diff} = .47\times10^{-12}$ [A] = .47 [pA].

The generation current is:

$I_{gen} = 1/2\; q\; n_i\; W\; A_j / \tau$;

$W = \sqrt{(2\, K_s\, \epsilon_o\, (\phi_B + V_R)) / (q\, N_B)}$

$= \sqrt{(2 \times 11.7 \times 8.86\times10^{-14} \times (.838+10)) / (10^{15} \times 1.6\times10^{-19})}$

$= 3.65\times10^{-4}$ [cm] = 3.65 [μm];

$\phi_B = (KT/q)\ln(N_B/n_i) + .55 = .288 + .55 = .838$ [V];

$$I_{gen} = \frac{1/2 \times 1.6\times10^{-19} \times 1.45\times10^{10} \times 3.65\times10^{-4} \times .25\times10^{-4}}{10\times10^{-6}};$$

$= 1.05\times10^{-11}$ [A] = 10.5 [pA].

The total current is: $I_R = I_{diff} + I_{gen} = 11$ [pA].

Exercise 6.1

Find the area of a p-n junction formed by diffusion using an oxide mask. The junction depth is 2.5 [μm], (lateral diffusion is 75% of vertical). The oxide mask dimensions are 500 [μm] by 500 [μm]. Assume cylindrical junction curvature at the edges and spherical at the corners.

Exercise 6.2

If the junction described above is abrupt and one sided (p^+-n) with the diffusion doped to 10^{19} [cm^{-3}] p type and the bulk being 2 [ohm cm] n type, find the diffusion and generation current components for a reverse bias of 15 [volts] and also the total reverse current. Approximate the carrier lifetime by using Figure 6.3.

The forward bias characteristic is given by:

$$I_F = I_R \left(e^{V_j/V_{Th}} - 1\right) \quad [\text{Amps}] \qquad (6.5)$$

Where V_j is the forward bias across the junction. I_R is computed with both reverse current components, using $V_R = 0$ in the formula for I_{gen}. The dependent current source to model current flow in a diode then is Equations 6.2 and 6.3 to find I_R, and Equation 6.5 to find the forward current, I_F. The model is shown below.

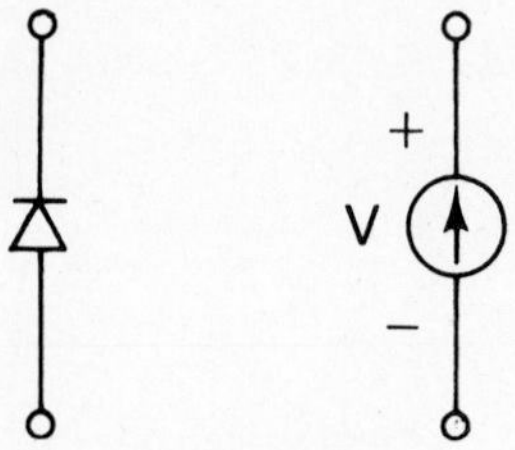

$$I = \begin{cases} I_F = I_R \left(e^{V/V_{th}} - 1\right) & ; V > 0 \\ 0 & ; V = 0 \\ -I_R & ; V < 0 \end{cases}$$

Exercise 6.3

Consider the junction of Exercise 6.2, find the forward current for forward biases of .2 [volts] and .5 [volts].

Exercise 6.4

Consider the junction of Example 6.1 and calculate the reverse current for every volt of reverse voltage and the forward current for every .1 [volt] of forward voltage for the range -10 [volts] $< V_j <$ 1 [volts]. Tabulate the results on graph paper and plot this graph. Enter the resulting graph in this workbook.

Exercise 6.5

An abrupt n^+-p junction of 400 [μm^2] of area, 2 [μm] junction depth fabricated on 2 [ohm cm] p type Si is measured to have a reverse bias current density of 10 [pA/cm^2] at 2 [volts] reverse bias. What is the minority carrier lifetime?

JUNCTION BREAKDOWN

In a reverse biased junction, the electric field can be considerable and exceed breakdown. By analyzing Poisson's equation for an abrupt one sided junction in one dimension, one may develop the following expression for the maximum field:

$$E_{MAX} = \frac{2 \times (V_R + \phi_B)}{\sqrt{\dfrac{2\, K_s\, \epsilon_o\, (V_R + \phi_B)}{q\, N_B}}} \quad [\mathrm{V/cm}] \qquad (6.6)$$

In intrinsic silicon, the maximum field before breakdown occurs is given in Table 1.1: $E_{crit} = 30\times10^4$ [V/cm]. As the doping increases, this breakdown field is increased. Plotted in Figure 6.7 we have E_{crit} versus doping. (There are two types of breakdown modes clearly labeled: avalanche and zener.) From Equation 6.6 one may solve for the breakdown voltage in terms of the critical field:

$$V_{BN} \cong (K_s\, \epsilon_o\, E_{crit}{}^2) / (2\, q\, N_B) \quad [\mathrm{Volts}] \qquad (6.7)$$

This gives us an indication of the breakdown of a planar junction. If there is any kind of curvature to this junction such as that shown in Figure 6.4, the breakdown field (and voltage) is reduced. A plot of this result for various curvatures is shown in Figure 6.6.

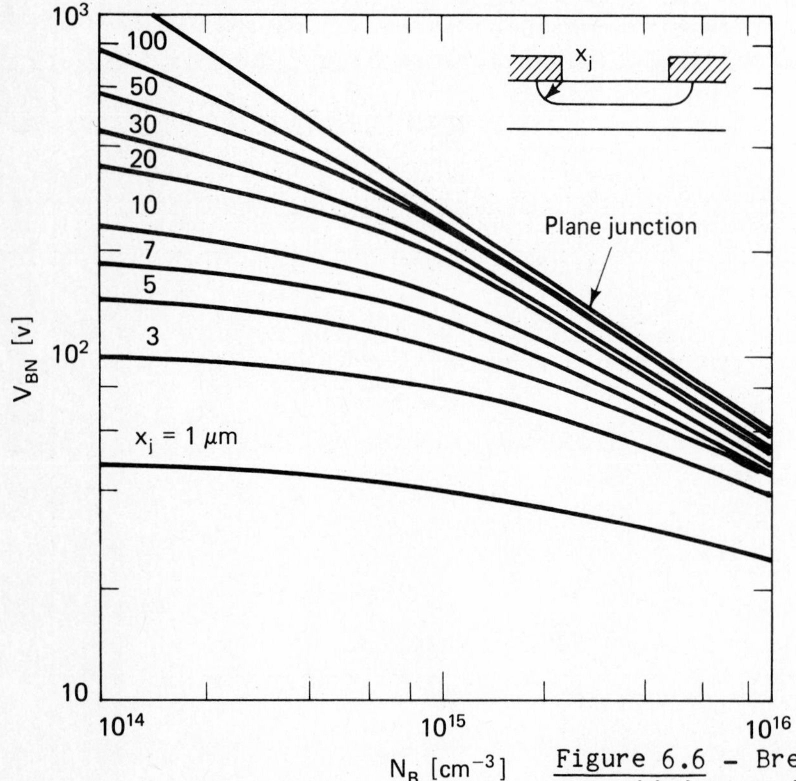

Figure 6.6 - Breakdown voltage of planar Si one-sided step junction showing effects of junction curvature. (After Armstrong. Reprinted with permission from IRE Transactions on Electron Devices, ED4, pg. 16, 1957.)

Exercise 6.6

Consider the junction of Exercise 6.4 and answer the following:

a. What is the critical field? What is the breakdown mode?

b. What is the breakdown voltage for a planar junction of similar junction depth?

c. Compute a corrected breakdown voltage to account for curvature.

d. If it is desired to make a planar zener diode with a 10 volt breakdown, what should the substrate doping be?

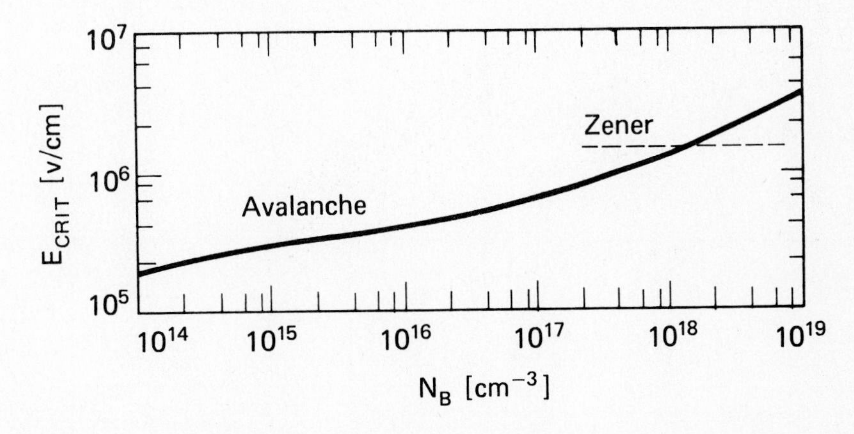

Figure 6.7 - Critical field for avalanche and zener junction breakdown in silicon. (After Grove. Reprinted with permission from John Wiley, Inc.)

Breakdown voltage and field are one of the limits of reverse bias operation of junctions. Therefore the model should be carefully constructed with the modeler in full awareness of these limits.

Computer Laboratory

Write and code an algorithm that computes forward or reverse current for an abrupt junction diode using the equations given. Your program should be flexible enough to handle $n^+ - p$ or $p^+ - n$ junctions and should have mobility functions built in so that the only inputs necessary are the applied bias and its polarity, doping of the bulk and required carrier lifetime. Assume silicon as the semiconductor. Solve Exercise 6.2 using the code as a check on your program.

CASE 7: DIODE

Objectives

1. Understand and develop a model of the capacitance of a p-n junction.

2. Develop and exercise the circuit model of a diode in a transient mode.

3. Analyze the switching characteristic of a p-n junction diode via a computer model.

Laboratory

Develop and code an algorithm that models the complete diode: capacitance, resistance and dependent current source, and exercises it for a transient switching example.

References

1. Glaser and Subak-Sharpe, Chapter 2.1.

2. Muller and Kamins, Chapter 4.4.

3. Grove, Chapter 6.

JUNCTION CAPACITANCE

Derivation of junction capacitance for a p-n junction diode is performed by solving Poisson's equation for the potential. The capacitance of a junction stems from the distribution of charges in the depletion region.

In reverse bias, the depletion width of an abrupt p-n junction is given by the following approximation:

$$W = \sqrt{2K_s\epsilon_o \ (N_A + N_D) \ (\phi_B + V_R)/q \ N_A N_D} \ \text{[cm]} \qquad (7.1)$$

The capacitance is expressed simply as the parallel plate capacitance:

$$C = (K_s\epsilon_o/W) \ \ A_j \ \ \text{[Farads]} \qquad (7.2)$$

In the case of an abrupt one-sided step junction we have:

$$W = \sqrt{2K_s\epsilon_o \ (\phi_B + V_R)/qN_B} \ \text{[cm]} \qquad (7.3)$$

All parameters are as defined in Case 6; the corresponding diagram is given in Figure 6.2.

The capacitance under forward bias may be quite considerable. It stems from the excess minority carriers "stored" in the junction under forward bias. They are "stored" because as they are injected across the depletion region, their presence above the background minority concentration constitutes a storage of charge. When the diode is suddenly switched from forward to reverse bias, this charge is momentarily trapped at the edge of the depletion region and is only then slowly reduced. Its presence constitutes a storage of charge and must so be accounted for as a capacitance:

$$C_{forward} = q/KT \ \tau \ I_D \ \text{[F]} \qquad (7.4)$$

where: τ is the minority carrier lifetime in the lightly doped side.

Exercise 7.1

Calculate and plot the capacitance per unit area for two abrupt one-sided n^+-p step diodes, manufactured on two different bulk resistivities 12 [ohm cm] and .1 [ohm cm]. Do this as a function of reverse bias.

		0	1	2	3	4	5	6	7	8
	Reverse bias [V]									
12 [ohm cm]	Depletion width [cm]									
	Capacitance [pF/cm^2]									
.1 [ohm cm]	Depletion width [cm]									
	Capacitance [pF/cm^2]									

Plot the above results for capacitance and include it in this workbook.

TRANSIENT DIODE MODEL

We shall now put together and exercise a diode model that includes a current source (see Case 6), and junction capacitance. The large signal equivalent circuit of the diode is shown below:

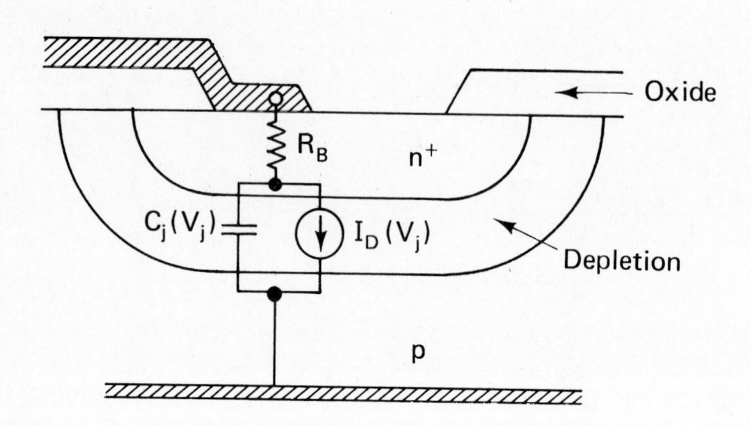

Figure 7.1 - Large signal model of a p-n junction diode.

Assume the bulk resistance to be negligible. The diode capacitance C_D is given by Equations 7.2 and 7.3 and the current source is given by the equation in Case 6. Let us analyze the step response of this diode. Consider the circuit in Figure 7.2.

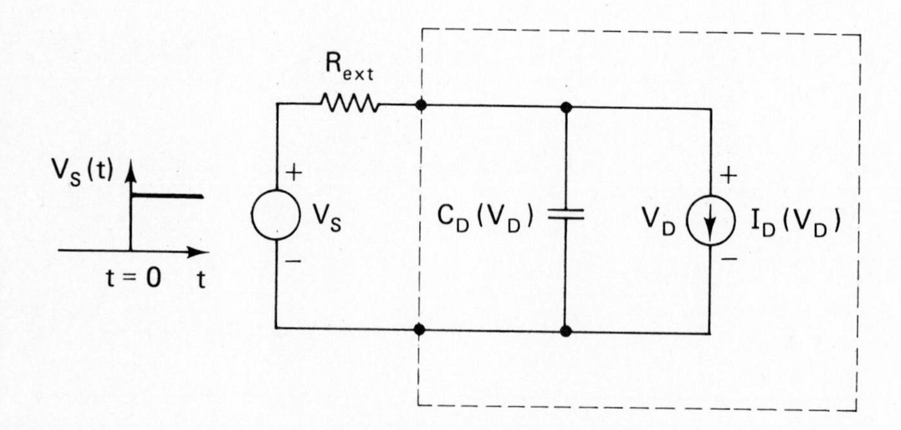

Figure 7.2 - Circuit for building a large signal computer model of a p-n junction diode.

We have then by circuit analysis:

$$dV_D/dt = I_C/C = (I_R - I_D)/C = ((V_S - V_D)/R - I_D)/C \qquad (7.5)$$

We approximate the derivative by a discretized time step:

$$V_D = ((V_S - V_D)/R - I_D)\Delta t/C \qquad (7.6)$$

The algorithm is given by Muller and Kamins in <u>Device Electronics for Integrated Ciruits</u>.

	Computer Operation	Parameter Values
1.	Initialize I_D, C, V_D for steady state with $V_S = 0$	$V_D = 0$, $I_D = 0$, $C = C_o$
2.	Set time t = 0	t = 0
3.	Set new V_S value	$V_S = -V_1$ for reverse bias
4.	Increment time	$t = t + \Delta t$
5.	Compute currents I_R and I_C	$I_R = V_S - V_D/R_S$ $I_C = I_R - I_D$
6.	Compute change in voltage ΔV_D during Δt	$\Delta V_D = I_C \quad t/C = (I_R - I_D) \quad t/C$
7.	Compute new diode voltage	$V_D = V_D + \Delta V_D$
8.	Calculate new diode current	$I_D = I_o(e^{qV_D/KT} - 1)$
9.	Test whether I_D has reached its steady state value: If YES >>>> EXIT If NO >>>> Continue	
10.	Calculate new depletion width and a new capacitance	W_D $C_D = K_S \epsilon_o A/W_D$ or $q/KT\tau I_D$
11.	Return to step 4.	

Part A - Implement and exercise the algorithm for the transient analysis of a diode. Use as an example an abrupt n^+-p diode, manufactured on 12 [ohm cm] substrate, 5 [μm] junction depth diffused through a 50 [μm] by 20 [μm] opening in a thick oxide mask. For a load resistance of 1000 [ohms] find and plot, as a function of time, the diode current, junction capacitance and diode junction voltage for a 10 [volt] step function as input.

Repeat for load resistances of 100 and 10,000 [ohms]. Add the results to this workbook.

Part B - Modify the algorithm above to accept input voltages as a function of time rather than just discrete time steps. Code and exercise the same problem as for part A above but use the following two time functions:

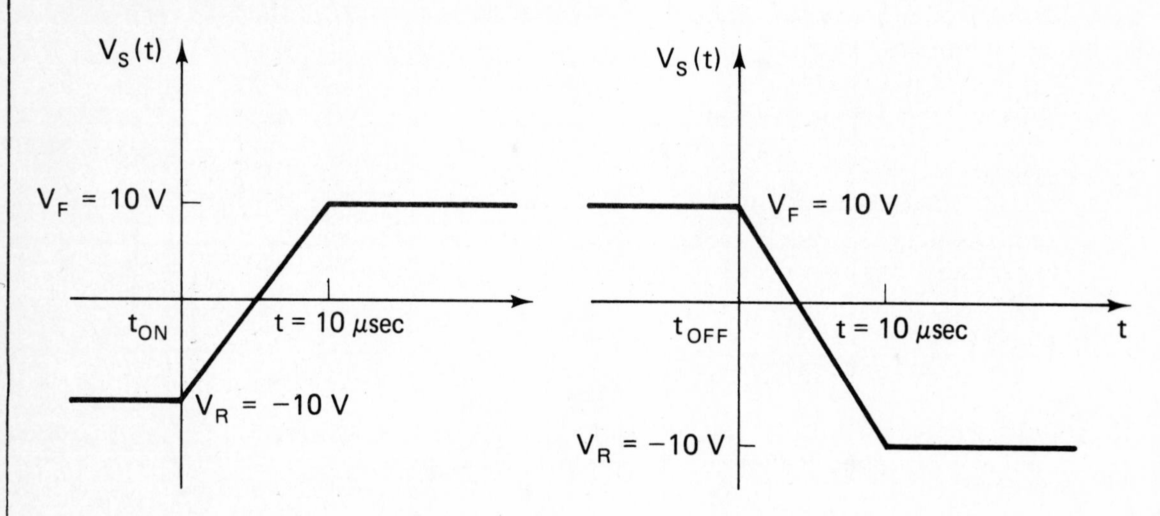

Add the results to this workbook.

CASE 8: PROCESSING

Objectives

1. Understand the process of diffusion of impurities in silicon.
2. Develop and manipulate a simple one dimensional model of diffusion.
3. Model the redistribution of typical profiles such as error function and gaussian.
4. Observe the effects of time and elevated temperatures on redistribution.

Laboratory

Develop a simple computer program to model one dimensional redistribution of deposited and diffused profiles.

References

1. Grove, Chapter 3.
2. Glasser and Subak-Sharpe, Chapter 5.4.
3. Gise and Blanchard, Chapters 9 and 10.

PRE-DEPOSITION

There are two important methods of introducing impurities into silicon. One is the traditional method of heating the wafer to elevated temperatures (1000°C–1200°C for Si) and allowing a gas of the impurity to flow over the wafer or having the material (as a solid) in contact with the wafer in the furnace. The second method which is much more controllable, done at room temperature and fast gaining acceptance for very large scale integration (VLSI), is ion implantation. In the present case we deal with traditional methods and in Case 10 we shall take up in more detail the redistribution of an ion implant profile. Keep in mind that we are seeking to understand and demonstrate basic principles and to develop and master simple models of these physical phenomena.

In high temperature pre-deposition, we introduce a controlled amount of impurity by controlling the time of pre-dep and the temperature at which we operate.

At a certain temperature only a certain maximum amount of an impurity may be "dissolved" in the substrate lattice. This amount is referred to as the "solid solubility" limit and it is plotted for various silicon dopants in Figure 8.1 as a function of temperature. This number fixes the <u>surface concentration</u> N_S.

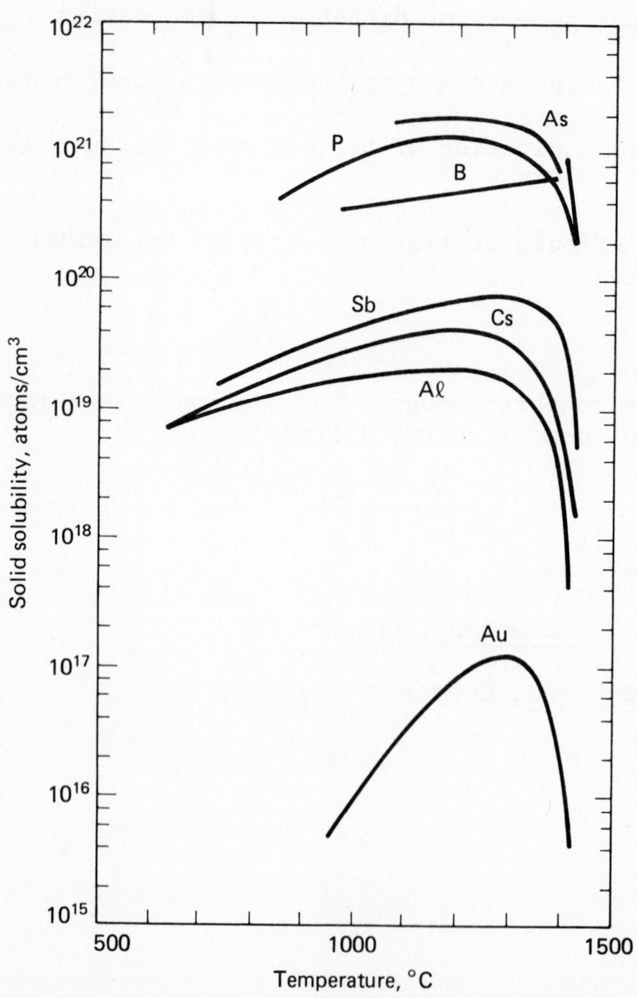

Figure 8.1 – Solid solubility of selected elements in silicon.

As the time of deposition is increased more material is absorbed by the silicon and the impurity segregates toward the bulk as is shown below.

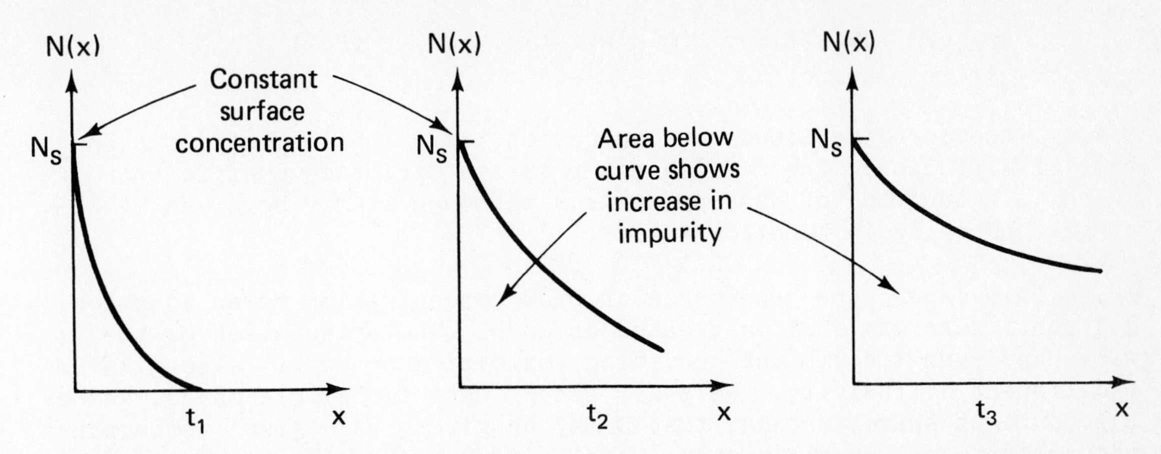

Figure 8.2 - Control of pre-deposition profile, changing surface concentration N_S, by changing temperature of pre-dep and changing time of pre-dep.

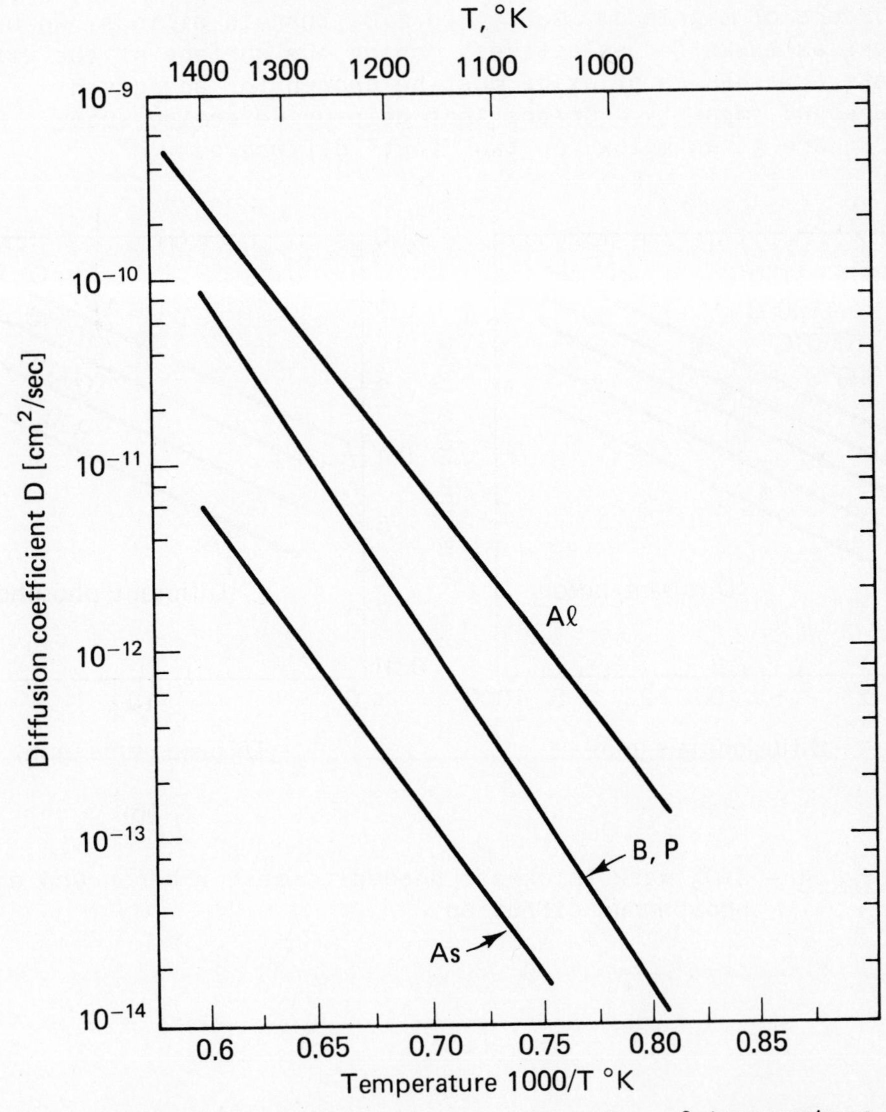

Figure 8.3 - Diffusion coefficient as a function of temperature for common diffusers in silicon.

The impurity ions diffuse into silicon in a characteristic manner described by a differential equation called Fick's Second Law:

$$dN(x,t)/dt = D\ d^2N(x,t)/dx^2. \qquad (8.1)$$

$N(x,t)$, the one-dimensional concentration, is a function of depth and time of diffusion. The coefficient D is the diffusion coefficient which is a function of many parameters although it may be given by a single number as in Equation 8.1.

In certain cases, the dependence of D on concentration makes Equation 8.1 non-linear and must be treated as such. The orientation of the wafer and even the ambient condition (oxidizing or inert) also give us a different diffusivity. We shall assume here for simple cases, and as a good first approximation, that D may be given, at a given temperature and ambient as a single number. A plot of D for various impurities in silicon is given in Figure 8.3.

Since the diffusivity of an impurity of interest (such as As, P or B) is many orders of magnitude smaller in SiO_2 than in silicon, we use SiO_2 layers as masks for selectively doping the surface of the wafer. Appropriate thicknesses of oxide must be chosen for any given temperature and impurity. Graphs that help us determine such thicknesses are given below for two "fast" diffusers.

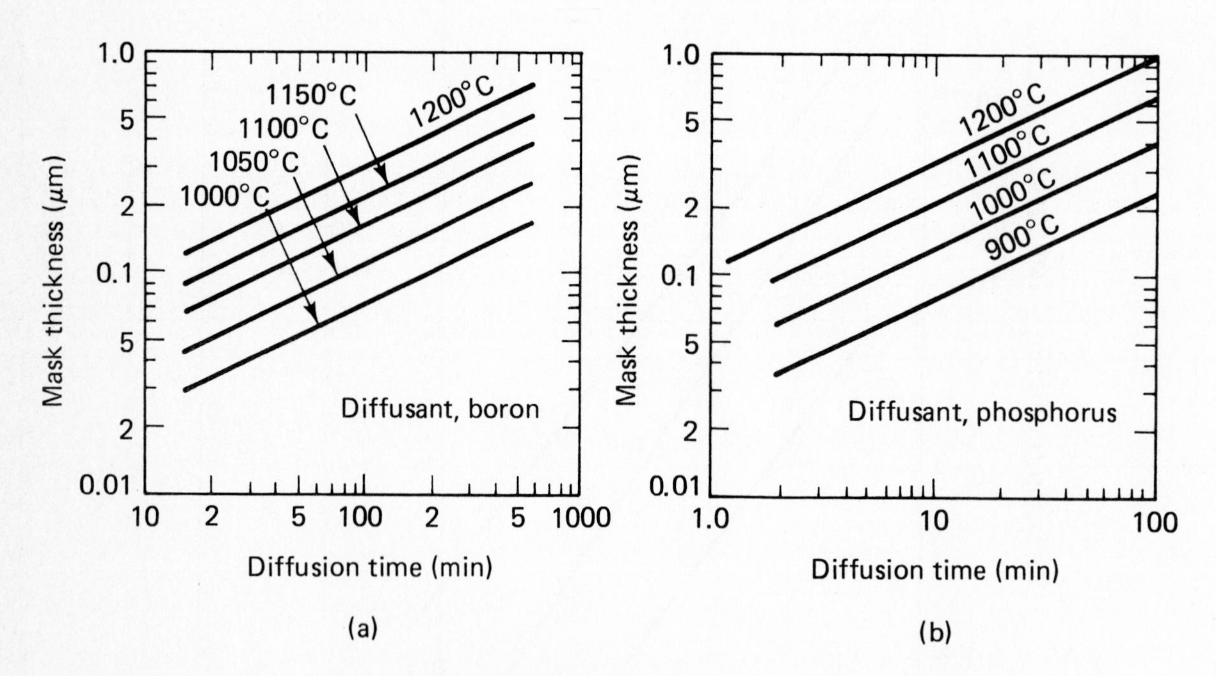

Figure 8.4 - SiO_2 mask thickness needed to mask a boron and a phosphorus diffusion.

Exercise 8.1

A p-n junction design calls for a phosphorus pre-deposition (at the solid solubility limit) into <100> p Si at $1000^{\circ}C$ for 100 minutes. Find:

a. Surface concentration after pre-dep.

b. Diffusivity of phosphorus at the pre-dep temperature.

c. SiO_2 mask thickness for masking this diffusion.

Repeat for a boron diffusion into <111> p Si at $900^{\circ}C$ for 120 minutes.

The solution of Equation 8.1 for the boundary condition of a solid or a gaseous pre-dep is given by:

$$N(x,t) = N_s \; \text{erfc} \; [x/ \sqrt{4D_p \; t_p} \;] \; [\text{ions/cm}^3] \qquad (8.2)$$

where: N_s = surface concentration $[cm^{-3}]$ (at the solid solubility limit for the pre-dep temperature);

x = depth in [cm] into the wafer measured from the surface;

D_p = diffusivity at the pre-dep temperature for the element in question in $[cm^2/sec]$;

t_p = time of pre-dep in [seconds].

The function erfc (·) is defined as:

$$\text{erfc}(z) = 1 - \text{erf}(z) = 1 - 2/\sqrt{\pi} \int_0^z e^{-y^2} dy \qquad (8.3)$$

For most purposes we shall use a tabulation of this function (see Appendix A) which is practical for simple problems. The integration is done numerically in the computer problem.

To find a given concentration $N(x,t)$ at a given depth x after a time t_p of pre-dep has elapsed, we must substitute this data into Equation 8.2. There are two items of interest besides the doping profile $N(x)$: the junction depth x_j (in the case of a p-n junction) achieved for a given pre-dep and the total amount of impurity introduced, Q, after pre-dep. The junction is that depth at which the impurity level introduced equals and cancels the background.

$$N_B/N_s = \text{erfc} \; [x_j/ \sqrt{4D_p t_p} \;] \qquad (8.4)$$

The junction depth is found by using the complementary error function table in reverse. The total amount of impurity introduced is given by:

$$Q = N_s \sqrt{4 \; D_p t_p / \pi} \quad [cm^{-2}] \qquad (8.5)$$

Example 8.1

Consider a pre-deposition step in the formation of a p-n junction. Phosphorus from a gaseous source is passed over a 12 [ohm cm] p Si wafer at 1000°C for 120 minutes. Calculate the junction depth.

Answer:

a. The surface concentration is found to be $N_s = 1.2 \times 10^{21}$ [cm^{-3}] of phosphorus using the graph in Figure 8.1.

b. The ratio $N_B/N_s = 1 \times 10^{15}/1.2 \times 10^{21} = 8 \times 10^{-7}$.

c. For erfc $Z = 8 \times 10^{-7}$ we find in the table of Appendix A that Z = 3.8. Also Z may be found using the approximate equation given in Case 10.

d. Then $x_j = \sqrt{4 D_p t_p}\ Z$, but $D_p = 3 \times 10^{-13}$ [cm^2/sec] for P in Si from Figure 8.3 and t_p = 120 [min] x 60 [sec/min] = 7200 [sec]:

$$x_j = \sqrt{4 \times 3 \times 10^{-13} \times 7200} \times 3.8 \text{ [cm]} = 3.13 \text{ [}\mu\text{m]}.$$

e. $Q = 1.2 \times 10^{21} \sqrt{4 \times 3 \times 10^{-13} \times 7200} / \pi = 3.55 \times 10^{16}$ [cm^{-3}].

Exercise 8.2

Consider a pre-dep of boron into a .1 [ohm cm] epitaxial layer of n type Si. The pre-dep time is 200 minutes at 1000°C. Find the junction depth and Q, the total charge introduced.

What thickness SiO_2 mask must be used to mask this pre-dep?

If it is desired to double the charge Q introduced but the temperature must remain constant, how long must the new pre-dep time be? Find the new junction depth.

DRIVE-IN

Once impurities are introduced and we shut off their source of supply and remain at high temperature, then any further redistribution still obeys Fick's Second Law (Equation 8.1) but now has a new solution:

$$N(x,t) = (Q/\sqrt{\pi D_2 t_2})\ \exp\,[-x^2/4D_2t_2]\ [\text{ions/cm}^3] \qquad (8.6)$$

Here we have made the assumption that the pre-dep profile is very narrow and may be modeled with a "delta function."

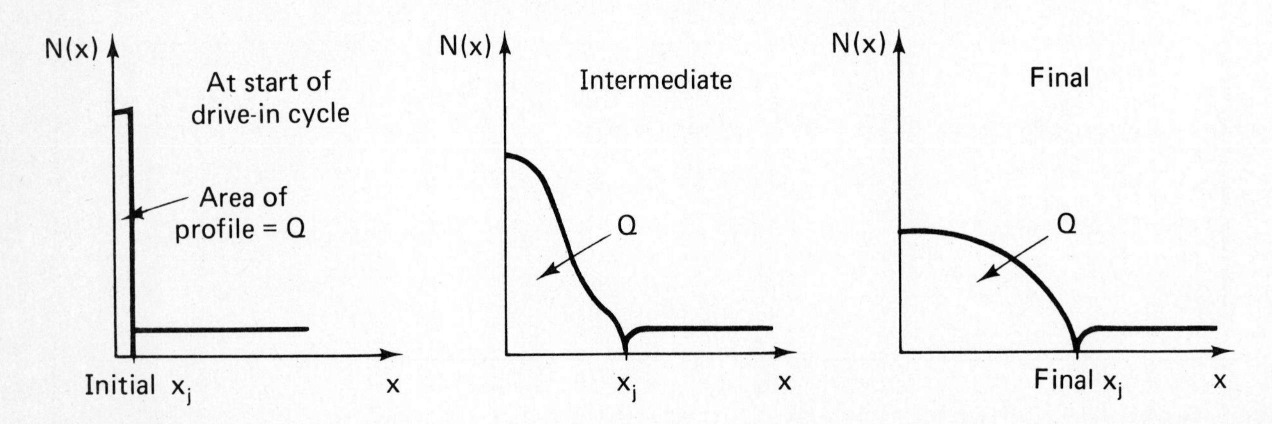

Figure 8.5 - Drive-in from a pre-deposited source concentration treated as a "delta function" shows the effect of redistribution and the gaussian profile for a junction.

This assumption is valid for any type of starting profile as long as it is "reasonably narrow" with respect to the final profile. Since the final profile has a straggle of $\sqrt{4D_d t_d}$, the Q approximation (Equation 8.6) is valid if:

a. for a pre-dep profile $x_j \ll \sqrt{4D_d t_d}$;

b. for an ion implanted profile $\sigma \ll \sqrt{4D_d t_d}$;

where: σ is the straggle as defined in Case 10.

The new junction depth is given by:

$$x_j = \sqrt{4D_2 t_2} \ \ln [Q/(N_B \times \sqrt{\pi D_d t_d})] \text{ [cm]} \qquad (8.7)$$

Exercise 8.3

The pre-dep given in Exercise 8.2 is followed with a drive-in cycle of 120 minutes at 1100°C. Compute the new surface concentration N(x=0) and junction depth.

Compare x_j from the pre-dep step to the appropriate drive-in parameter and justify the use of Equation 8.6

Use Equation 8.6 and compute the final profile N(x) after drive-in for every .05 [µm] from x=0 to $x=x_j+.5$ [µm]. Make a table below and plot the results; include them in the workbook.

Exercise 8.4

Consider ion implanting arsenic into a 12 [ohm cm] p Si substrate. The implant parameters are such that a dose (Q) of 4×10^{15} [cm^{-3}] with a straggle (σ) of .05 [µm] results in the silicon after implant (considered as the pre-dep in this case). Find the final junction depth after a drive-in cycle of 120 minutes at 950°C.

Write and code a computer program that will analyze a process sequence in the following manner:

Part A (pre-dep) - Program accepts diffusivity, time at temperature and substrate doping. It computes:

a. The profile using Equation 8.2 given depth increment and desired maximum depth.

b. The total charge Q.

The program should utilize a subroutine to integrate the error function and compute erfc (Z) given Z. This routine is similar to that written for Case 10.

Part B (drive-in) - This part may be entered separately from Part A but may also be a continuation of Part A. It computes the profile and junction depth after a drive cycle as follows:

a. Receives Q from Part A or external input and, if following a pre-dep step, also accepts x_j; if following an implant one needs to input σ.

b. Accepts D_2 and t_2 for the drive-in cycle, computes $\sqrt{4D_2t_2}^{\,2}$ and checks the validity of the approximation behind Equation 8.6.

c. If the equation is valid, program continues by computing and printing the profile for the given increment and depth information.

d. Computes and prints junction depth if appropriate (i.e., if a p-n junction has been formed).

Validate the program by recomputing the solutions to Exercises 8.2, 8.3, and 8.4.

Exercise 8.5

A certain $PClO_3$ pre-dep/drive-in cycle used to manufacture a junction for a VLSI chip is defined as follows:

a. Pre-dep - 10 min. @ 1000_oC

b. Drive-in - 120 min. @ 1100°C

The substrate is 12 [ohm cm] p Si.

a. Find the final junction depth.

b. Find, tabulate, and plot the final profile.

c. It is desired to decrease the junction depth by 50 percent but keep the surface concentration constant. What would you change? By how much? Do it and design the new profile. It is necesary to solve Equation 8.7 for t_2. Note that a transcendental equation for t_2 results. It is most easily solved iteratively by a simple computer program.

d. Use the computer program that calculates sheet rho (from Case 2) and compute the sheet rho of both the original and redesigned junctions for this exercise.

CASE 9: PROCESSING

Objectives

1. Understand the process of oxidation of silicon.

2. Compute the resulting oxide thicknesses from dry and wet oxidation processes, as a single step or for multiple events.

3. Understand the effect of oxidation on impurity distribution.

4. Develop more accurate models of diffusion in the presence of oxide boundaries.

Laboratory

Implement a more accurate and general model for diffusion and include it in the code developed in Case 8.

References

1. Grove, Chapters 2 and 3.

2. A. B. Glaser and Subak-Sharpe, Chapter 5.

3. Gise and Blanchard, chapters 7 and 8.

4. E. Douglas and A. Dingwall, "Ion Implantation for Threshold Control in COSMOS Circuits", IEEE, TED-21, No. 6, June 1974, pp. 324-331.

OXIDATION

There are two basic processes for oxidation broadly termed dry and wet. Conventional dry oxidation is a much slower oxide growth process than wet oxidation. For the same amount of time and at the same temperature, the wet oxide is many times thicker than the dry oxide. Experimentally determined oxide thicknesses for Si are shown in Figures 9.1 and 9.2. These will be the workhorse graphs for determining oxide thicknesses.

Example 9.1

Determine the oxide thickness for a 180 [min] $1000^{o}C$ wet oxide cycle starting from a bare wafer.

Answer: Using Figure 9.2 we read 2 [μm] or 20,000Å of oxide.

Example 9.2

It is desired to mask a 20 minute $900^{o}C$ phosphorus diffusion with an SiO_2 mask. Design the wet process cycle required to give the proper mask thickness.

Answer: From Figure 8.4, determine t_{ox} = 1.0 [μm] then going to Figure 9.2, choose $1000^{o}C$ and read 70 [min].

A factor of importance in choosing heat cycle parameters is the cumulative and individual impact they have on the redistribution of impurities. A measure of this redistribution is the D·t product, diffusivity times time at temperature. To compare two processes one compares the $\sqrt{D \cdot t}$ (which is in units of length) for each process.

Example 9.3

Which process introduces greater redistribution of impurities: 60 [min] @ $1000^{o}C$ or 10 [min @ $1100^{o}C$?

Answer:

For the $1000^{o}C$ process, t = 3600 [sec], D = 3×10^{-14} [cm^2/sec];

$\sqrt{D \cdot t} = 3 \times 10^{-14} \times 3600$ = .1 [μm].

For the $1100^{o}C$ process, t = 600 [sec], D = 3×10^{-13} [cm^2/sec];

$\sqrt{D \cdot t} = 3 \times 10^{-13} \times 600$ = .13 [μm].

The $1100^{o}C$ process then produces greater redistribution.

Exercise 9.1

At a certain point in a process it is necessary to mask a 1150°C boron diffusion which takes place for 70 [min]. It is also required to cause the least amount of redistribution in the profile. Choose a heat cycle to manufacture the required oxide mask. (HINT: answer the questions below.)

a. Find the required mask thickness:

b. Choose either a wet cycle or a dry cycle necesary to grow the oxide for part (a) and justify your choice.

c. Use Figures 8.3 and 9.1 or 9.2 (as indicated) and fill in the following table:

	Temperature [°C]			
	1200°	1100°	1000°	900°
Read diffusivity [cm²/sec]				
Read time [sec]				
Calculate $\sqrt{D \cdot t}$ [μm]				

Choose from the process cycle that causes the least redistribution (minimum $\sqrt{D \cdot t}$). This is the design that satisfies both requirements.

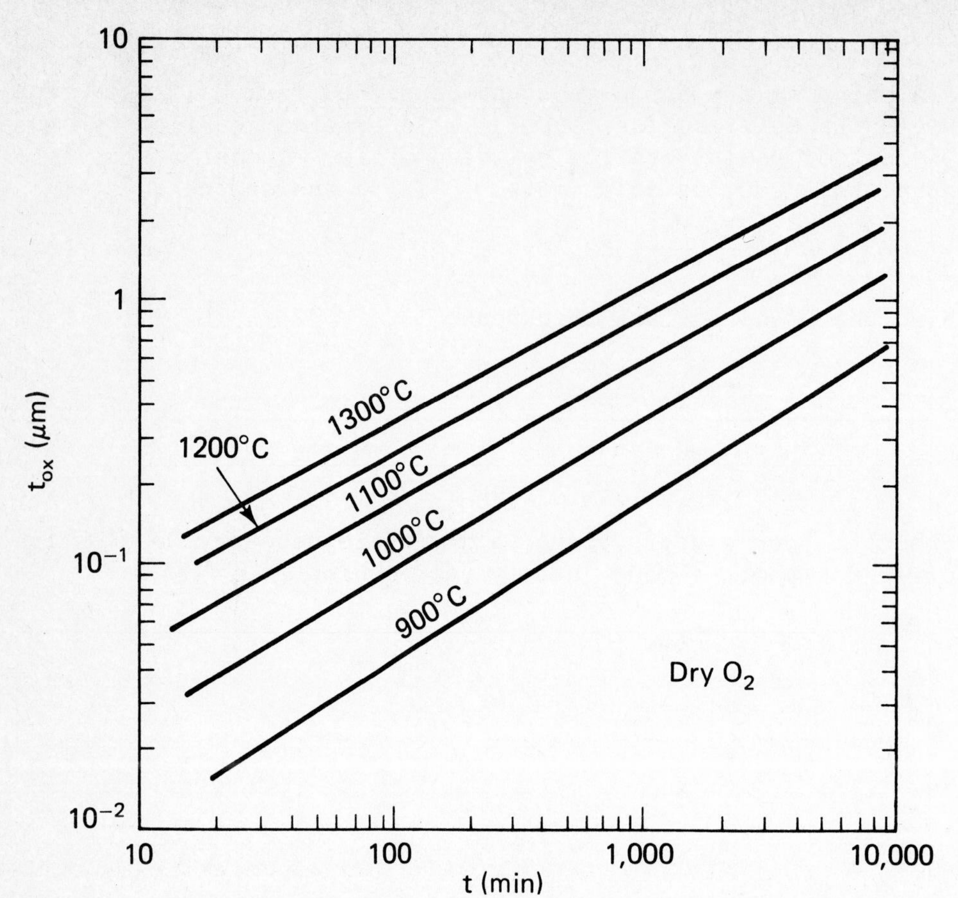

Figure 9.1 - Thermal oxidation of Si as a function of time in dry oxygen.

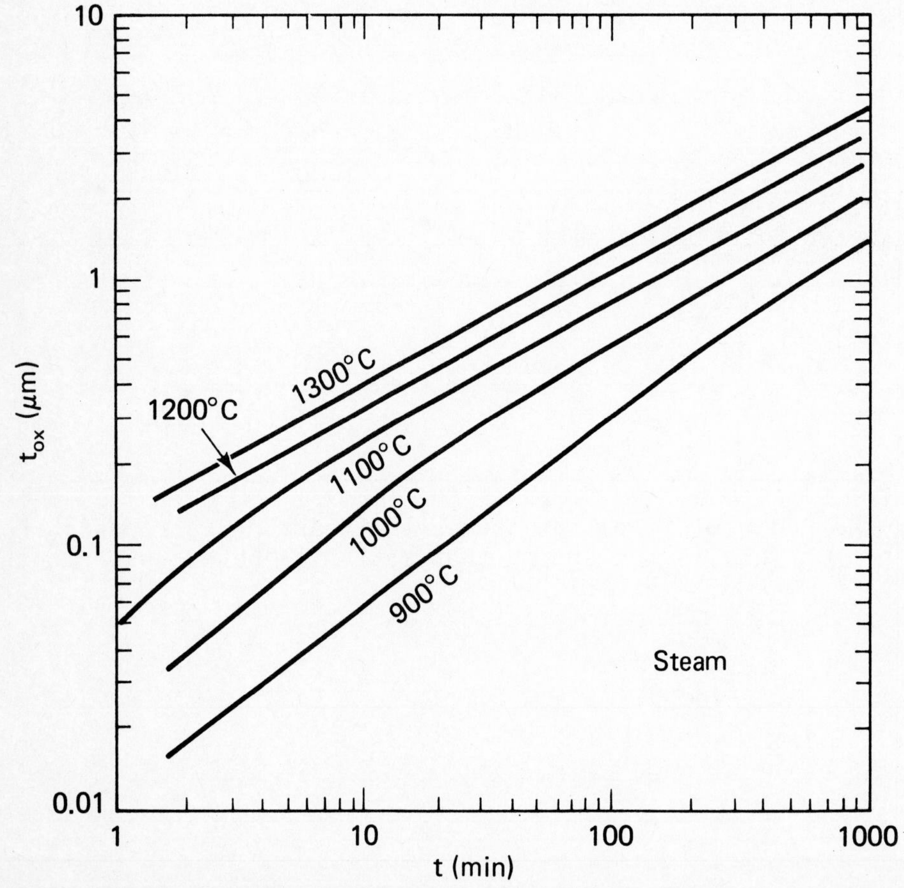

Figure 9.2 - Thermal oxidation of Si as a function of time in steam.

Oxide does not grow uniformly as a function of time. Look at equal increments of time in either Figure 9.1 or 9.2. They don't yield equal increments in oxide thickness. We may nevertheless analyze a series of oxidation steps using both graphs by the principle of equivalent oxide thickness. By this we mean that the oxide already on the wafer, whether produced by one or many oxidation steps may be considered as if it had already been produced in the present oxidation step and the new time is just added to the equivalent time at the present temperature.

Example 9.4

An oxide of 1,000Å is already on a wafer. Compute the oxide thickness after 3 hours at $1000^{o}C$ dry (O_2).

Answer: 1,000Å (.1 [µm]) of oxide could have been produced at $1000^{o}C$ dry and it would have taken an equivalent 2 hours (120 min.) to produce. To find the new oxide, add 180 minutes to the equivalent 120 minutes for a total of 300 minutes at $1000^{o}C$ dry to obtain t_{ox} = 1,800Å.

Exercise 9.2

Compute the total oxide thickness for the following process sequence:

Cycle	Temperature [^{o}C]	Time [min]
1. Wet O_2	1000	150
2. Dry O_2	1000	120
3. Wet O_2	950	50

This is a typical process history for the field oxide region of a VLSI n-channel poly-gate MOS process and may be analyzed as follows:

a. Compute the oxide thickness for wet O_2 $1000^{o}C$ 150 [min] process.

b. Convert this oxide thickness to an equivalent time at $1000^{o}C$ dry, add to the given time for cycle 2 and compute the total oxide.

c. Convert this new oxide to an equivalent time at $950^{o}C$ wet and add to the given time and re-compute. This is the final oxide thickness.

d. Now suppose 2,000Å of oxide were etched off between cycles 2 and 3 (this frequently happens in a process). Re-compute the final oxide thickness.

e. Compute the complete heat cycle history step by step by tabulating $\sqrt{D \cdot t}$ for each step, and the total by computing:

$$\sqrt{Dt} = \sqrt{D_1 t_1 + D_2 t_2 + \ldots\ldots} \qquad (9.1)$$

In this case for all three process steps the result is:

f. Which heat cycle has the greatest impact?

Exercise 9.3

Consider the process described in Exercise 9.2. It has been found that the second oxidation cycle is to be performed for 90 minutes rather than 120 minutes. It is necessary that the final oxide be the same as before but the total "time at temperature" be constant (i.e., $\sqrt{Dt}$ be kept constant). Design a change in the 950°C oxide step so the oxide thickness comes out properly and then add a drive-in cycle between cycles 2 and 3 to compensate and keep $\sqrt{Dt}$ constant.

When the oxide of silicon grows, it does so by consuming silicon as the O_2 molecules interact with the Si atoms. If there had been impurities in the Si, then depending on the type and their diffusivity in the oxide they will either be depleted (in the case of boron) or pile up at the Si-SiO_2 interface (as in the case of phosphorus). The diagrams below show this process.

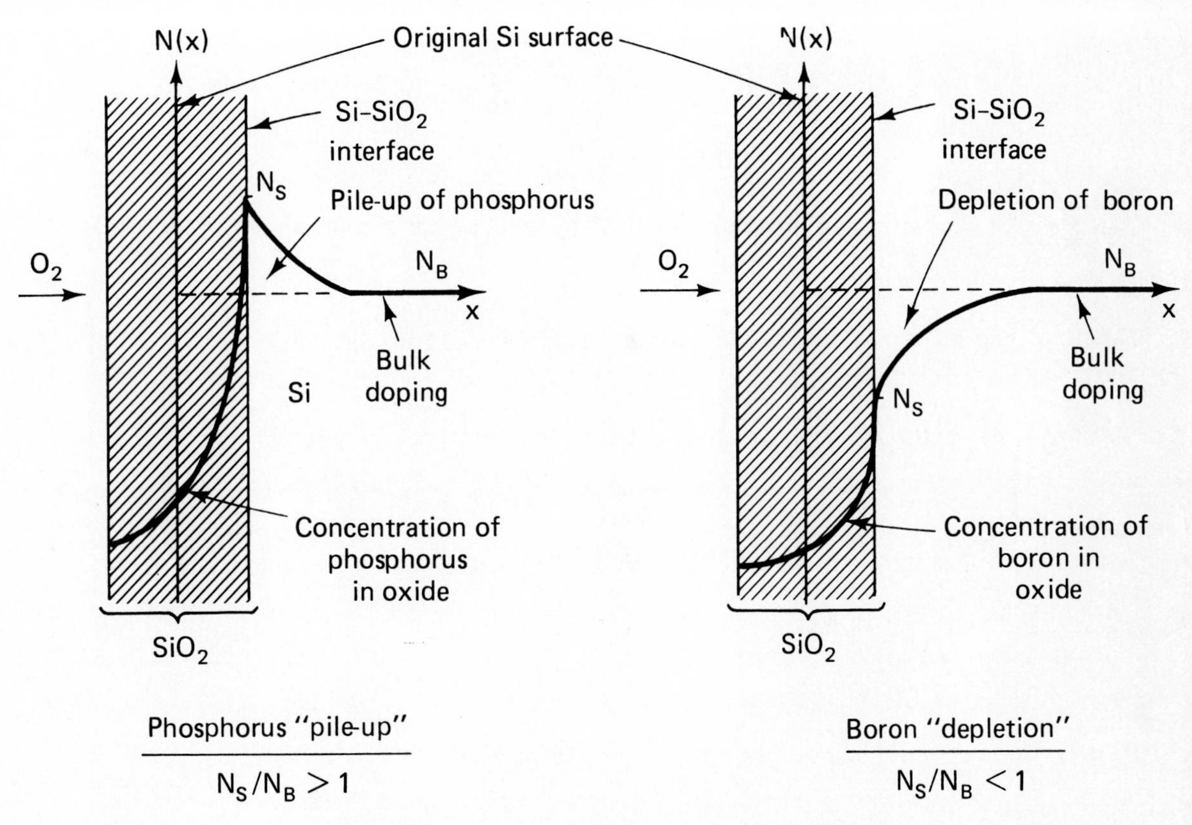

Figure 9.3 - Redistribution of phosphorous and boron at an interface during thermal oxidaton.

This pile-up or depletion is considerably removed when a drive-in cycle of significant consequence is introduced and impurities redistribute.

DIFFUSION MODEL

A more exact model of diffusion for gaussian profiles is discussed next. Consider a profile that may be described by a gaussian function (see Case 4):

$$N(x) = N_{po} \exp\left[-(x - \mu)^2 / 2\sigma^2\right] + N_B \ [\mathrm{cm}^{-3}] \qquad (2.4)$$

Consider that part of this profile may reside in the oxide as shown in the diagram below:

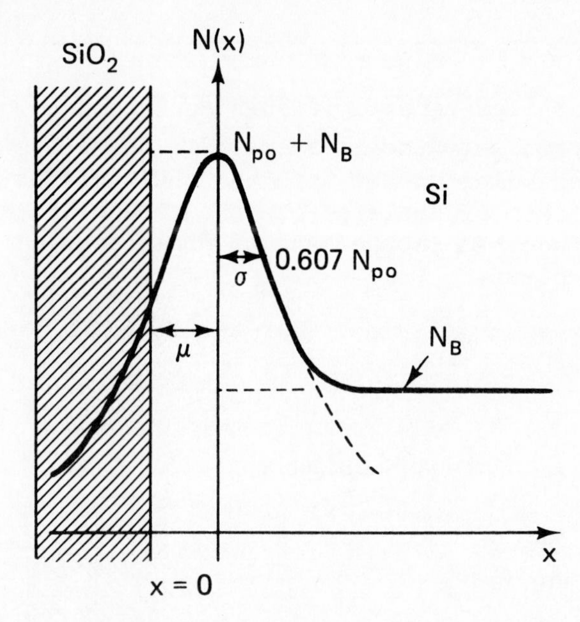

Figure 9.4 - Gaussian profile defining parameters and showing profile in SiO_2 as well as Si.

The profile after a drive-in cycle has the following form:

$$N_{Si}(x) = \left(N_{po}/2\sqrt{\pi Dt}\right)\left[O(x,t) + O(-x,t)\right] + N_B \ [\mathrm{cm}^{-3}] \qquad (9.2)$$

where:
$O(x,t) = \sqrt{C}\exp\left[-(A - B^2/4C)\right] \ (\sqrt{\pi}/2)\ (1 + \mathrm{erf}\left[(x - B/2)/\sqrt{C}\right]);$
and:

$A = (x-\mu)^2/2\sigma^2;$
$B = 4Dt\ (x-\mu)/(\sigma^2+2Dt);$
$C = 4Dt\ \sigma^2/\ \sigma^2+Dt\ .$

In this case, the total charge Q is given by:

$$Q = \sqrt{\pi}/2 \quad N_{po}\ \sigma \quad (1 + \mathrm{erf}\left[\ \mu/\sqrt{2\sigma}\right])\ [\mathrm{ions/cm}^2]. \qquad (9.3)$$

Computer Laboratory

Develop a new version of the diffusion model of Case 8 which shall be more accurate for drive-in of gaussian profiles by implementing the equations above. Change the algorithm of Case 8 in the following way:

1. Accept only gaussian parameters: N_p, μ , σ , N_B.

2. No high temperature pre-dep is allowed.

3. Replace the drive-in solution (Equation 8.5) with Equation 9.2.

4. Replace the equation for Q with Equation 9.3.

5. No junction calculation is necesary.

6. Retain the error function integration routine as it is needed in this case.

Validate this program using the data and results of Exercise 8.5.

This last program is most useful to study redistribution of ion implants for MOS threshold adjust (see Case 11).

CASE 10: PROCESSING

Objectives

1. Understand ion implantation as a pre-deposition step with its associated parameters.

2. Model the processing of a threshold adjust ion implant for a typical n channel poly-gate process.

3. Design implant conditions given a process sequence and final desired profile.

Laboratory

Exercise a simple process model by studying the redistribution of an ion implanted profile.

References

1. G. Dearnaley, J. H. Freeman, R. S. Nelson, and J. Stephen, Chapter 5.

2. Glaser and Subak-Sharpe, Chapter 5.5.

3. W.S. Johnson and J.F. Gibbons, Projected Range Statistics in Semiconductors, Stanford University Press, 1969.

4. P. H. Rose and A.B., Whittkower, "An Introduction to Ion Implantation Equipment", Varian Technical Report VR-102, Varian Associates, Palo Alto, Ca., 1975.

ION IMPLANTATION

Ion implantation is a relatively low temperature method of introducing impurities into a substrate. A schematic of a typical machine used in producing ionized beams of dopant ions, of great purity, accelerating, then scanning them onto a sample is shown in Figure 10.1. In a typical VLSI process, there may be as many as six different ion implant steps. Among those applications which require design, study and modeling are: threshold adjusts (low dose, low energy) and source-drain implants (high dose) for MOSFET technologies; and base implants (low dose, varying energy) and emitter junction and epitaxial doping (high to medium doses) for bipolar processes.

Two implant parameters must be ascertained in a design: the energy of the beam E [KeV] and the total dose D [ions/cm^2]. The penetration of the beam and its initial spread are dependent on beam energy. The total charge introduced is dependent on the dose. Since in many instances one wishes to dope the substrate selectively, then care must be taken to insure that the masking medium provides sufficient protection from penetration at the energy employed.

The profile immediately after an ion implant step is considered to be gaussian as a first approximation. The profile has the following expression:

$$N(x) = N_{Rp} \exp - [(x-R_p)^2 / 2(\Delta R_p)^2] + N_B \text{ [cm}^{-3}] \qquad (10.1)$$

where: N_{Rp} = the peak concentration determined by dose [cm^{-3}];

R_p = the range of penetration determined by ion type and energy [μm];

ΔR_p = straggle of the distribution determined by ion type and energy [μm];

N_B = bulk doping [cm^{-3}].

This is similar to Equation 4.1 but the symbols here are consistent with those used in ion implantation. The range R_p is the depth of penetration of the implant, dependent on ion energy and type. The straggle ΔR_p is a measure of the spread of the implant, also dependent on energy and ion type, and is equivalent to the symbol σ of Equation 4.1. Both of these parameters are printed in Appendix B as a function of energy and for the three most common ion types.

This doping profile, once introduced, may be thought of as a pre-depositon and treated by the techniques of Case 9 when considering redistribution. A typical profile is drawn in Figure 10.2. For data on penetration into materials other than silicon and for other ions the reader is directed to Reference 10.3.

The relationship between peak concentration and dose (the area under the profile) is found from the integral of the gaussian profile:

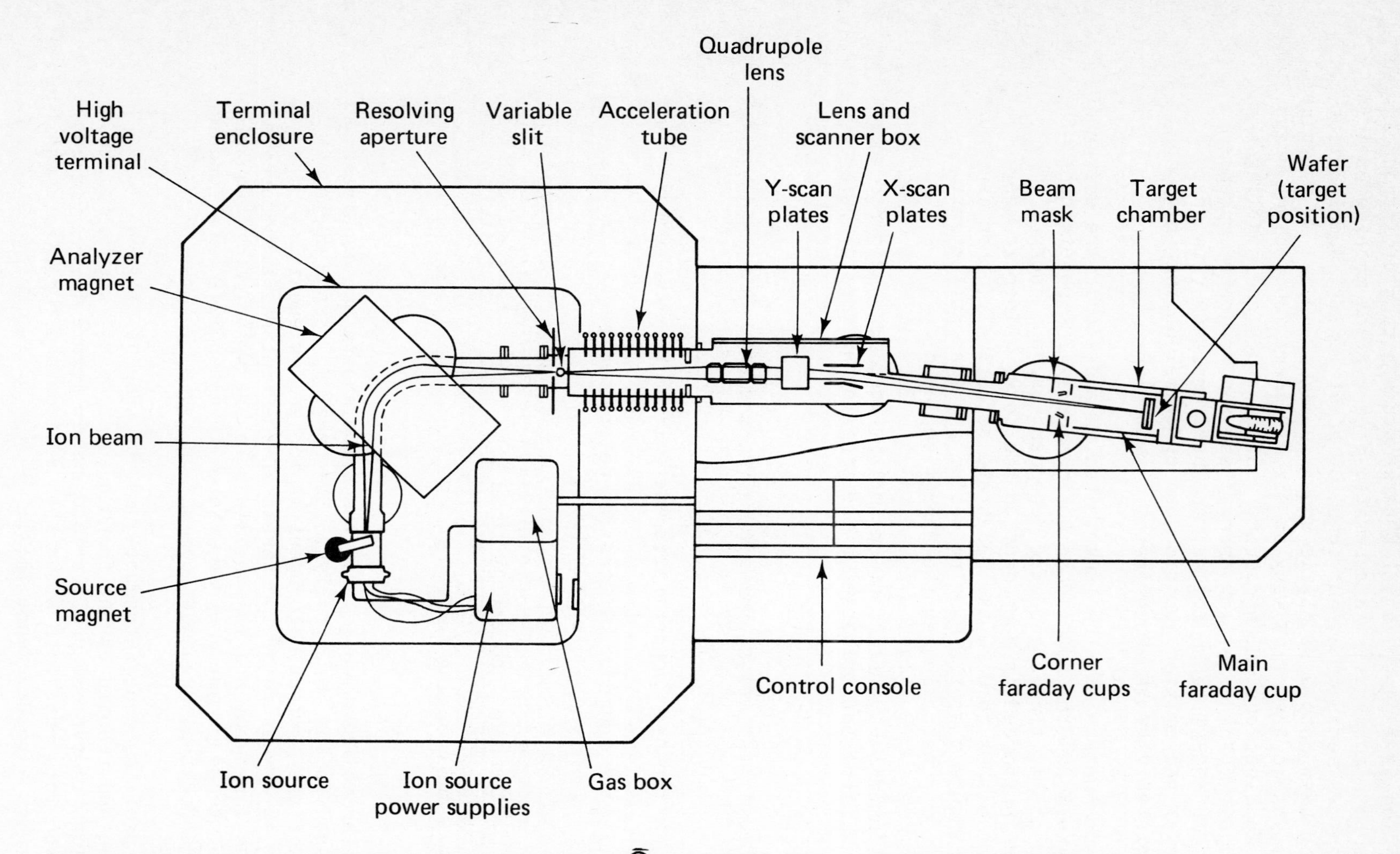

Figure 10.1 - Top view of a Varian Extrion® ion implanting machine (200 DF4). This is typical of the medium current ion implanters which are currently the workhorse of the industry. This machine will implant typically between 25 and 200 [KeV] and accepts wafers up to 4 [in]. (Reprinted with permission from Varian Associates/Extrion Division, Gloucester, MA., 01930, USA.)

$$\text{Dose} = N_{Rp}\, \Delta R_p \sqrt{2\pi}\ [\text{ions/cm}^2] \qquad (10.2)$$

Example 10.1

Compute the depth of penetration of a 100 [KeV] B^+ beam into silicon and the dose necessary to have a peak of 1×10^{18} [ions/cm^3] immediately after the implant step; the bulk doping is 1×10^{17} [cm^{-3}].

Answer: For a 100 [KeV] B^+ beam implanted into silicon and using the tables in Appendix B one finds:

$R_p = .3[\mu m]$, $\Delta R_p = .062\ [\mu m]$;

$N_{Rp} = N_{peak} - N_B = .9\times10^{18}$ [ions/cm^3];

$N(x) = .9\times10^{18} \exp -[(x-.3)^2/(.062)^2]$;

then: $\text{Dose} = N_{Rp}\, \Delta R_p \sqrt{2\pi}$;

$\text{Dose} = .9\times10^{18} \times .062\times10^{-4} \times \sqrt{2\pi} = 1.44\times10^{13}$ [ions/cm^2].

Exercise 10.1

Consider comparing two n type ion beams of different dopant type: As^+ and P^+. Find the straggle and depth of penetration of each ion into silicon for a 150 [KeV] beam.

	R_p [μm]	ΔR_p [μm]	N_{Rp} [cm^{-3}]
As^+			
P^+			
B^+			

Complete the table by obtaining the same parameters for a boron beam of the same energy. Also compute N_{Rp} for all three ions at this energy for a dose of 1×10^{12} [ions/cm^2]. Enter results into the table above.

In many cases the implant is done through a thin insulating layer. In these cases not all the implanted ions are resident in the substrate and it is important to know how much charge has been transmitted into the substrate. There are basically four cases (see Figure 10.3):

Case 1. The penetration P is much greater than the mask thickness, $R_p - P > 3\Delta R_p$ in which case almost all of the implanted dose equals the charge in the substrate. This case may be treated as in Example 10.1.

Case 2. P is greater than the mask thickness but less than $3\Delta R_p$, in which case the charge is partially given by the dose. Its calculation is done below.

Case 3. P is negative, i.e., the peak of the profile lies within the mask, but there is a considerable tail, $P < 3\Delta R_p$. This case is also treated below in detail.

Case 4. The mask is thick enough to mask the profile completely, $P > 3\Delta R_p$.

Computing the amount of charge implanted into a substrate requires calculations involving the error function discussed in Case 9.

Case 1 - Calculation of the charge from the dose in this case is simply given by:

$$Q = \text{Dose} \qquad (10.3)$$

Case 3 - It is easier to treat Case 3 before Case 2. Here we are trying to find the area under a normal function.

$$Q = \int_T^0 N(x)\, dx, \qquad (10.4)$$

then: $Q = \sqrt{2\pi}/2 \; N_{Rp} \Delta R_p \; \text{erfc}\; [(T - R_p)/\Delta R_p]$,

and: $Q = \text{Dose}/2 \;\; \text{erfc}\; [(T - R_p)/\Delta R_p]$,

the error function is tabulated in Appendix A.

For the special case where $T = R_p$, then:

$$Q = \text{Dose}/2 \qquad (10.5)$$

Case 2 - In this case we may subtract the part of the dose which lies in the mask from the total dose. Observe that this portion is not implanted into the substrate and may be obtained in the same manner as the portion implanted in Case 3, thus:

$$Q = \text{Dose}\ (1 - \text{erfc}\ [(R_p - T)/\Delta R_p]) \qquad (10.6)$$

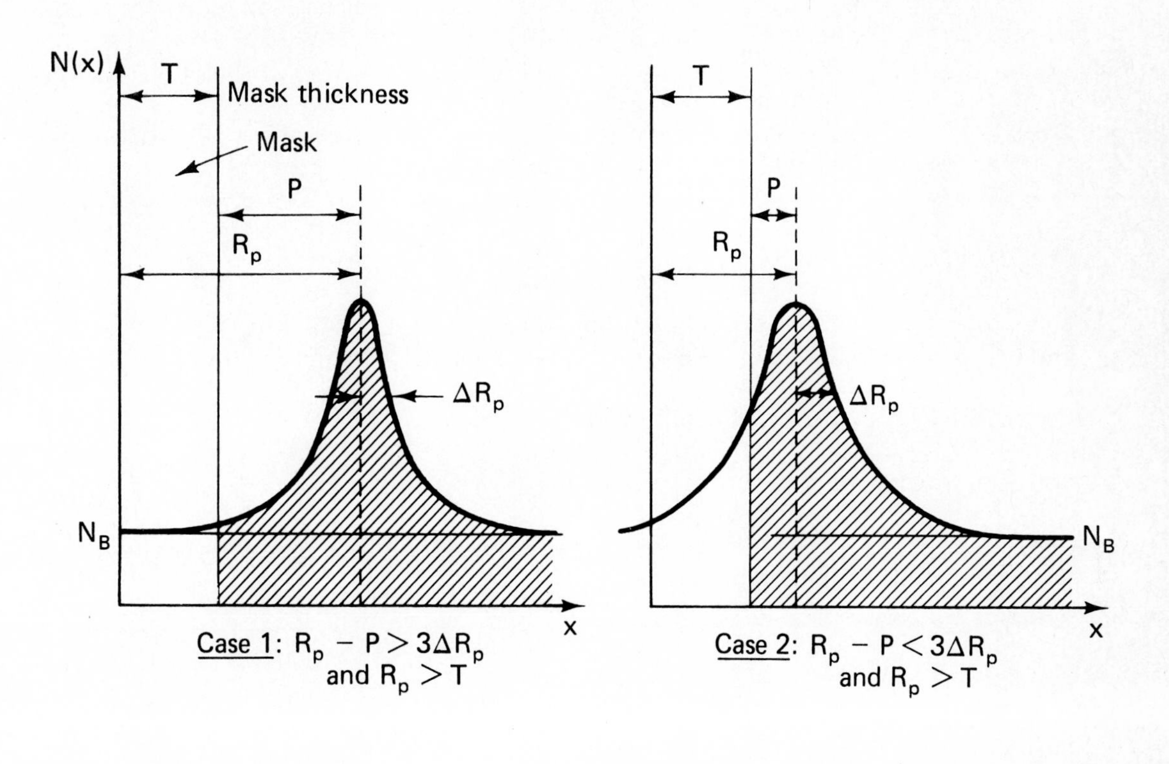

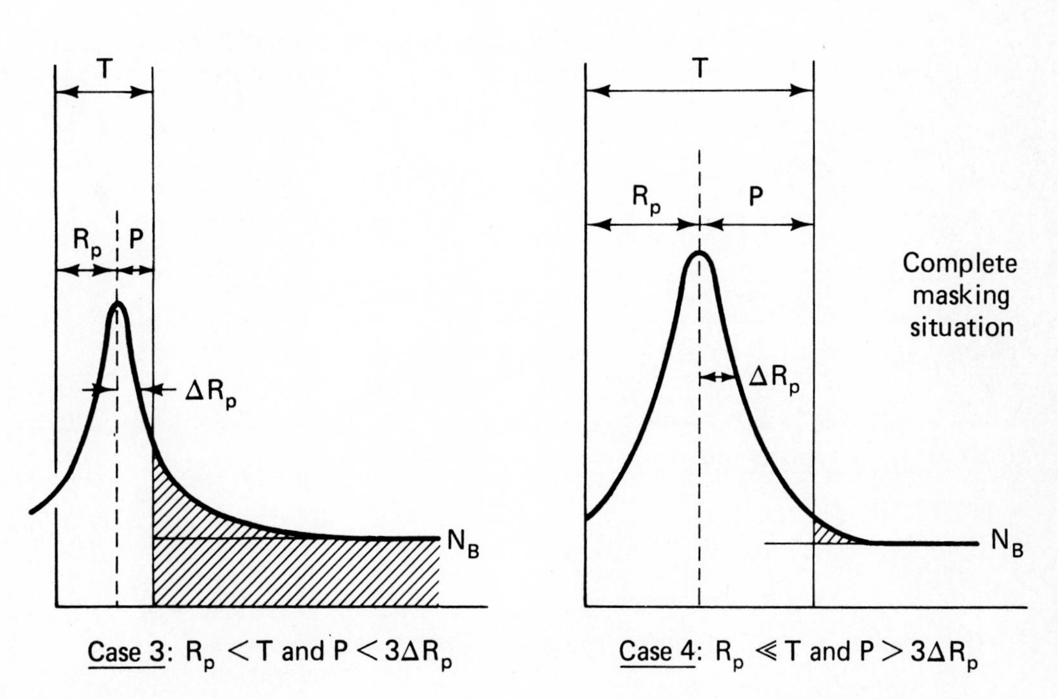

Figure 10.2 - The four possible cases that arise when ion implanting through a screen.

Example 10.2

Consider implanting B^+ at 100 [KeV] through a 4000Å SiO_2 mask. Calculate the percentage transmission of doping ions into a silicon substrate.

Answer: The stopping power of SiO_2 is approximately the same as that of silicon, thus we may consider both materials as one as far as beam penetration is concerned. For B^+ at 100 [KeV] we have $R_p = .3$ [μm] and $\Delta R_p = .062$ [μm]. Then since $R_p < T$ and $R_p - T = -P = .1$ [um] $< 3\Delta R_p$, we have the conditions of Case 3.

$$Q = \int_{.50}^{\infty} N(x)\,dx = (\text{Dose}/2)\ \text{erfc}\ [(.4-.3)/.062)],$$

$$Q/\text{Dose} = \text{erfc}\ [1.6]\ /2 = .23/2 = .110 = 11\%.$$

If the mask thickness is reduced to 2000A then $T < R_p$ but $R_p - T = P = .1$ [μm]$< 3\Delta R_p$ and we have the elements of case 2.

$$Q = \text{Dose}\ (1 - \text{erf}\ [(.3-.2)/.062])\ \text{and}$$

$$Q/\text{Dose} = (1 -.23) = 77\%$$

We see then that for the reduced screen thickness, a greater percentage of the dose penetrates the screen.

Exercise 10.2

Compute and tabulate the percent transmission for a P^+ ion beam into Si through a 4000Å SiO_2 "screen" (or mask) for two energies 50 [KeV] and 100 [KeV].

Phosphorus P^+

Energy	Case type	R_p [μm]	ΔR_p [μm]	Q/Dose (%)
50 [KeV]				
100 [KeV]				

Example 10.3

Consider an 800A Si_3N_4 mask over Si. Compute the implanted charge into the silicon for a 10^{12} [cm^{-2}] dose of B^+ at 50 [KeV].

Answer: The stopping power of Si_3N_4 is approximately .75x that of silicon. Thus for B^+ @ 50 [KeV], (R_p = .16 [μm] and ΔR_p = .04 [μm]) the Si_3N_4 thickness may be converted into an equivalent Si thickness by multiplying by the ratio of stopping powers (this may be roughly obtained from the ratio of the range for both materials at this energy and for this ion, .75 as stated).

T_{eq} = .75 x (800Å) = 600Å = .06 [μm].

Then, since $T < R_p$ and $R_p - T < 3\Delta R_p$ we have Case 3:

Q = Dose/2 erfc [(.2 - .06/.07)] = 10^{12} x .74 = 7.4 x 10^{11} [ions/cm^2].

Exercise 10.3

A certain two layer mask is present over silicon during a boron implant. Compute the necessary energy of the B^+ beam and the dose such that the peak concentration of 5.6×10^{17} [ions/cm^3] is obtained and a total charge of 10^{13} [ions/cm^2] is present in the silicon.

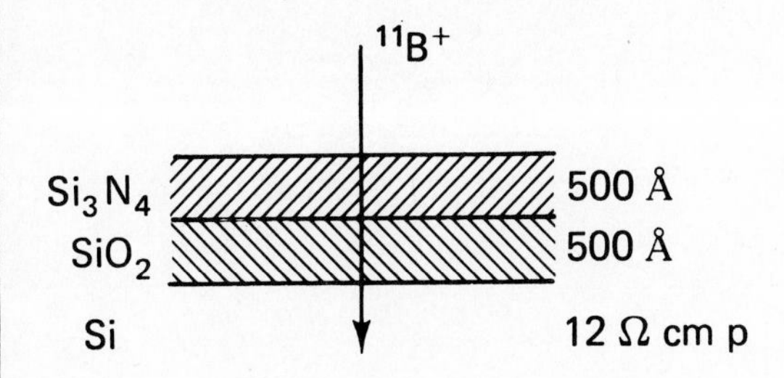

Example 10.4

Compute the range of any given ion beam in terms of its straggle such that for a given mask thickness T less than .1% of the dose penetrates into the substrate.

Answer: This is Case 4 but we shall analyze it as Case 3 to obtain the required relationship.

$Q = Dose/2 \ \ erfc\,[(T - R_p)/\Delta R_p];$

$Q/Dose = .01 = erfc\,[(T - R_p)/\Delta R_p];$

this occurs for: $T - R_p/\Delta R_p = 1.16;$

and: $T - R_p = 1.16\,\Delta R_p.$

Actually to act as an efficient mask, one must insure that $T - R_p = 3\,\Delta R_p$ as a minimum thickness.

Exercise 10.4

Compute a mask thickness such that less than .05% of a B^+ beam reaches the substrate at 50 [KeV].

Computer Laboratory

Write and code a computer program to analyze ion implant situations. Inputs to this program must be range and straggle of the beam and mask thickness if any is present as well as the substrate doping. It should compute any of the following parameters with the appropriate inputs for others as required.

Compute	Given
N_{Rp}	Dose
Dose	N_{Rp}
Q, Q/Dose	Dose or N_{Rp}

Also it should print out a sufficient number of points to enable the plot of the function N(x) to be made. Run this program for the conditions given in Exercise 10.2 as a validation check. Use either a computer accessible routine for the erfc(z) or write your own routine using the algorithm below.

Generate a set of tables for B^+, As^+ and P^+ of the percent transmission versus energy for implanting through SiO_2 masks of various thickness (500Å, 750Å, 1000Å, 5000Å) into a Si substrate. Plot three graphs, one for each ion, with beam energy as the ordinate and percent transmission as the abcissa and curves for each oxide thickness plotted simultaneously. The energy should range from 25 [KeV] to 200 [KeV], the typical range of most ion implanters..

USEFUL APPROXIMATION TO THE ERROR FUNCTION

For computer implementation of the error function the following rational approximation is given by Abromowitz and Stegun in Handbook of Mathematical Functions, Dover, N.Y., page 299:

$$\mathrm{erf}(x) = 1 - (a_1 t + a_2 t^2 + a_3 t^3)\ e^{-x^2}$$

where $t = 1/(1+px)$,

$p = .47047$,

$a_1 = .3480242$,

$a_2 = -.0958798$,

$a_3 = .7478556$.

CASE 11: MOSFET

Objectives

1. Understand MOSFET drain current versus drain voltage characteristics and the parameters that affect them.

2. Employ active region current models in device modeling.

3. Develop a threshold model from an active current region extrapolation.

Laboratory

Develop and code an algorithm that models the active current MOSFET characteristics.

References

1. Richman, Chapter 4.

2. Glaser and Subak-Sharpe, Chapter 3.4.

3. Sze, Chapter 10.

DRAIN CURRENT - DRAIN VOLTAGE CHARACTERISTIC

When a voltage greater than threshold is applied to the gate of the MOSFET, then a channel is formed and there appears at the surface a sufficient number of carriers for a current to flow when a drain to source voltage is applied. There are two regions of operations above threshold. One is the active region, the other: saturation. Figure 11.1 shows typical I_{DS} - V_{DS} characteristics with the corresponding channel conditions for each region of operation sketched in.

The current below saturation is given by:

$$I_{DS} = \beta \left\{ V_{DS} (V_{GS} - V_{FB} - 2\,\phi_B - V_{DS}/2) - 2K/3\,((V_{DS} + 2\,\phi_B)^{3/2} - (2\,\phi_B)^{3/2}) \right\} \text{[Amps]} \qquad (11.2)$$

where: $K = \sqrt{2 \quad (\epsilon_s N_A)} \;/\; (C_{ox}/A)$,

and $\beta = \epsilon_{ox}\, \mu\, W / T_{ox}\, L$.

A very good approximation to this current, below saturation, is given by:

$$I_{DS} = \beta \left\{ (V_{GS} - V_T)\, V_{DS} - V_{DS}^{3/2} \right\} \text{[Amps]} \qquad (11.2)$$

Saturation of the drain current occurs when the channel is pinched off. This occurs when the drain to channel electric field is stronger than the gate to channel field. The channel region nearest the drain becomes completely depleted and the drain current becomes a diffusion current.

Saturation occurs at a drain to source voltage given by:

$$V_{DS} \cong V_G - V_T \text{ [volts]} \qquad (11.3)$$

and the saturation current is:

$$I_{DSAT} \cong \beta/2 \quad (V_{GS} - V_T)^2 \text{ [Amps]} \qquad (11.4)$$

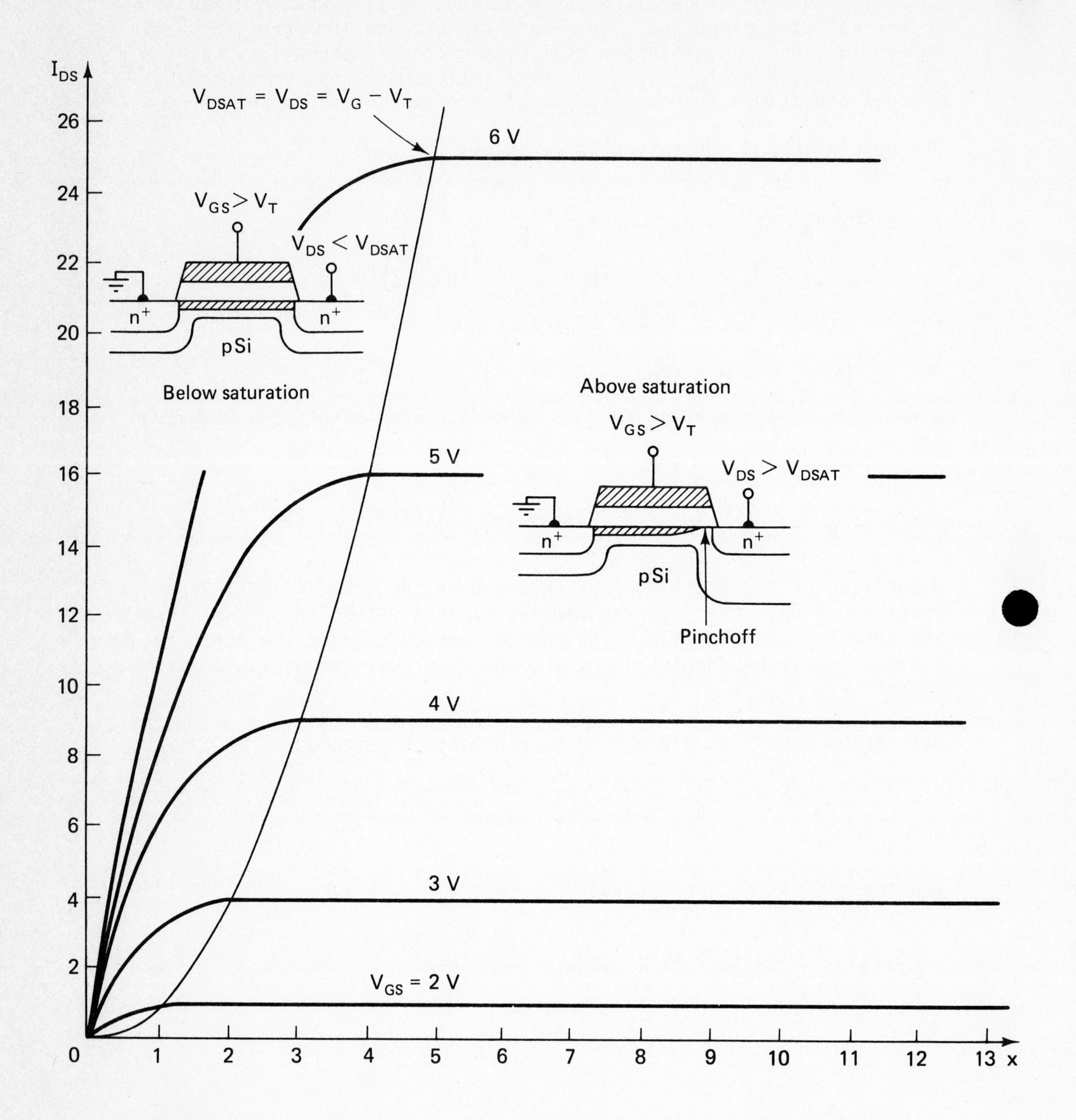

Figure 11.1 - MOSFET drain current characteristics above and below saturation.

Exercise 11.1

Consider an n channel aluminum gate MOSFET fabricated on 2 [ohm cm] p Si, $Q_{SS} = 0$. The oxide thickness is 1000Å, W/L = 10.

a. Find the threshold voltage at zero substrate bias.

b. Find the saturation voltage and current for $V_{GS} = 10V$.

c. Re-compute (a) and (b) for a device with a 2000Å oxide.

d. Re-compute (a) and (b) with a 1000Å oxide but for a p channel device on 2 [ohm cm] Si.

One may obtain threshold (V_T) from I_{DS} - V_{DS} characteristics by extracting from the curves values of V_{GS} versus $\sqrt{I_{DS}}$ for a fixed value of $V_{DS} > V_{SAT}$. If this is plotted, a straight line extrapolation of the points yields an ordinate (I_{DS} = 0) intercept which gives V_T. This is indeed the gate voltage for which the drain current is zero.

Example 11.1

For the MOSFET characteristics given in Figure 11.1, find the threshold voltage if the device is built on .1 [ohm cm] p Si, T_{OX} = 1500Å. Find Q_{SS} (assume V_{SS} = 0 [V].)

Answer Choose V_{DS} = 6 [V] for the five characteristics plotted. This is well above saturation for all of them.

V_{GS}	2 [V]	3 [V]	4 [V]	5 [V]	6 [V]
$\sqrt{I_{DSAT}}$	1 [A]	2 [A]	3 [A]	4 [A]	5 [A]

This is plotted in Figure 11.2 where a linear extrapolation shows the intercept to be V_T = 1 [volt].

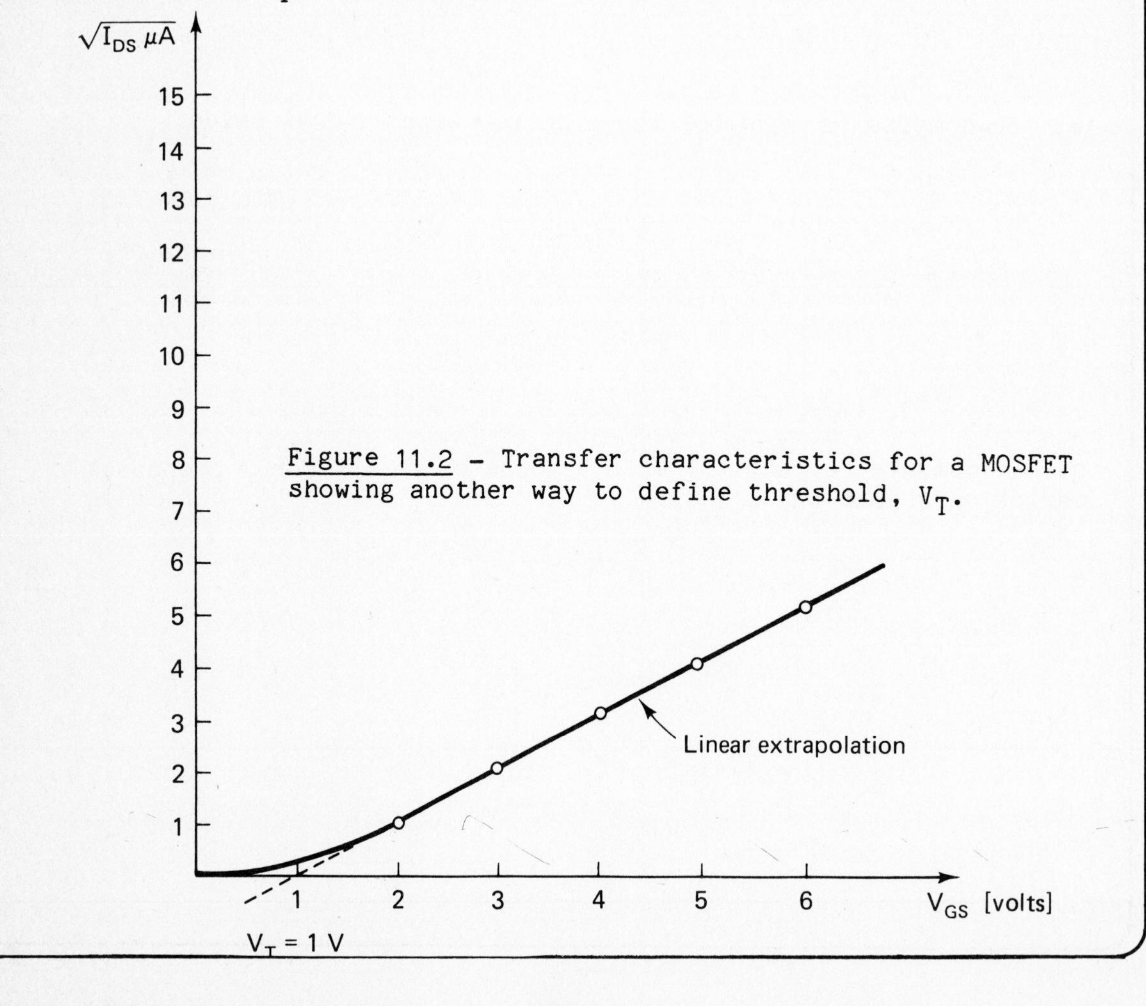

Figure 11.2 - Transfer characteristics for a MOSFET showing another way to define threshold, V_T.

Exercise 11.2

Consider the measured I_{DS} - V_{DS} characteristics given in Figure 11.3. This MOSFET is built on 2 [ohm cm] p Si with a 430Å oxide gate insulator, metal gate, W = 28.9 [μm] and L = 19.1 [μm]. Using the technique demonstrated above, find V_T and Q_{SS} for this device and compare both answers to that found in Exercise 5.2 (it is the same device).

NOTE: $\sqrt{I_{DS}}$ - V_{GS} data should be plotted for extrapolation purposes and the resulting graph added to this workbook.

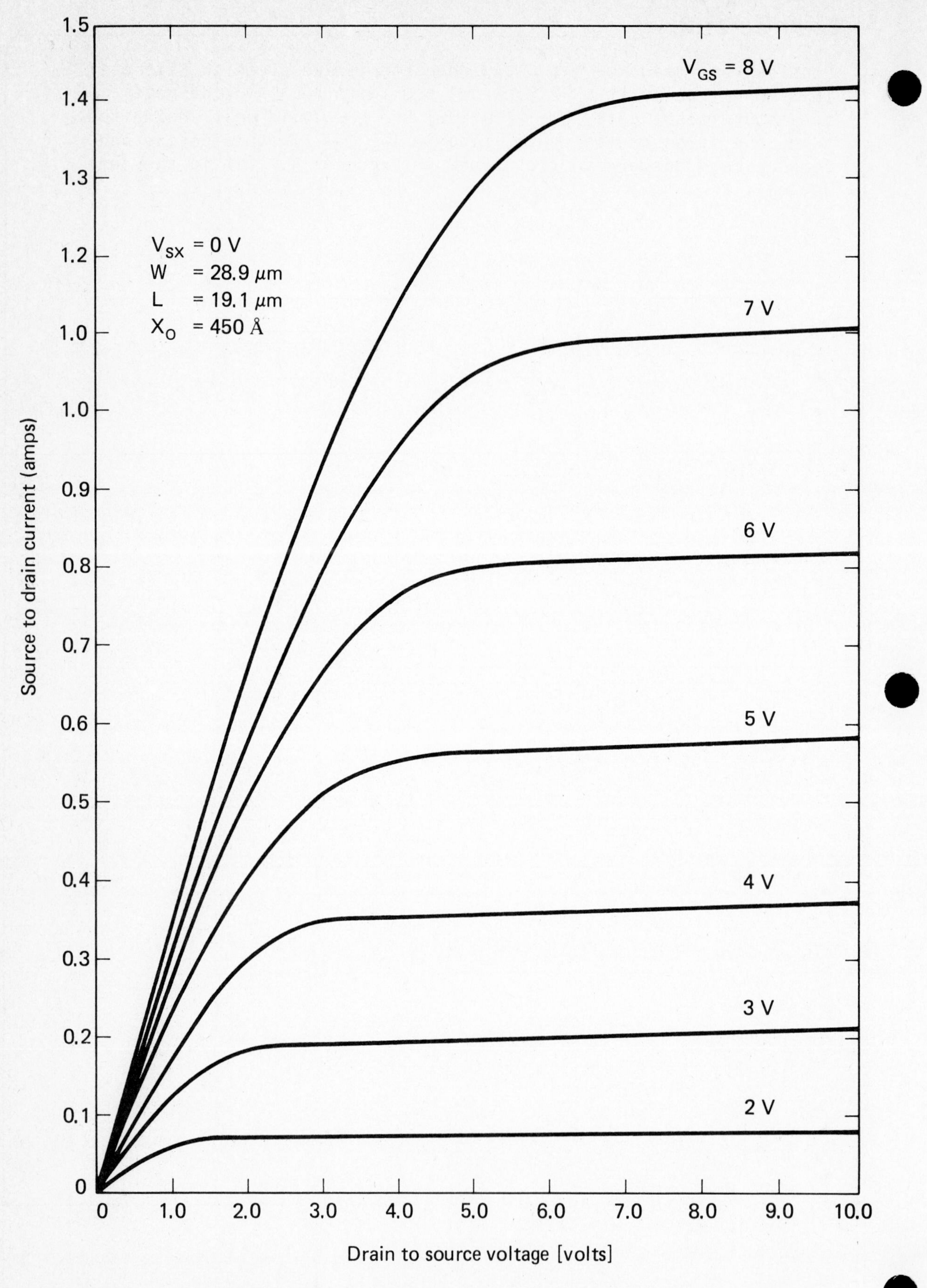

Figure 11.3 - MOSFET drain characteristics for Exercise 11.2. Channel length is 19.1 [um], channel width is 28.9 [um] and the oxide thickness is 430 Å. Measurements were made at V_{SS} = 0 [V].

The use of bulk mobility values in calculating the drain current yields results incompatible with experimental data. Since the carriers travel very close to the surface and there is a great deal more scattering in this region, the mobility there is very much reduced and a smaller current results. This surface mobility may be ascertained from measured I_{DS} - V_{GS} data by measuring I_{DSAT} and using Equation 11.4, solving for the mobility.

$$\mu_S = (2\ T_{ox}\ L)/(\epsilon_{ox}\ W)\ [I_{DSAT}/(V_{GS}-V_T)^2]\ [cm^2/V\ sec] \qquad (11.5)$$

Before a model may be used, it should be checked with data to prove its accuracy and reveal its weaknesses. It is often necesary to refine a model created on first principles (on just the basic physics) with experimental data. Such is the case with the surface mobility computed in this case.

Exercise 11.3

Consider once again the I_{DS} - V_{GS} data of Figure 11.3. Calculate the mobility of this device by matching the drain current in saturation to that computed using Equation 11.4. Average at least five values.

Compare the results to the bulk mobility for 2 [ohm cm] p Si. Account for any discrepancy.

Exercise 11.4

This exercise involves using the model Equations 11.2 and 11.4. Use the given data for the device and that computed in Exercises 11.1 and 11.2. Compute I_{DS} for various values of V_{DS} at V_{GS} = 5 volts and fill in the table below.

V_{DS} [V]	.5	1.0	1.5	2.0	2.5	3.0	4.0	5.0	6.5	8.0
I_{DS} [μA]										

Compute V_{DSAT}.

Plot these points superimposed on the data of Figure 11.2 and compare measured and modeled curves. Account for any discrepancy.

Computer Laboratory

Write and code a computer program that models the current characteristics of a MOSFET. It should have as its input:

1. Channel mobility;
2. Bulk type and doping;
3. Threshold voltage;
4. Insulator dielectric constant and thickness;
5. Range of drain bias.

The program should generate a table of drain current versus drain voltages for various gate voltages, i.e., a characteristic plot. Validate the program by generating all the characteristics given in Figure 11.3 and plotting them superimposed on the given data.

CASE 12: CIRCUIT ANALYSIS

Objectives

1. Understand and derive the parameters of a MOSFET circuit model for DC and transient analysis.

2. Investigate inverter fundamentals and analyze DC transfer characteristics of various inverters.

3. Understand the transient operation of an inverter and capacitive loading.

Laboratory

Develop a computer program for DC and transient modeling of inverters.

References

1. Glaser and Subak-Sharpe.

2. Till and Luxon, Chapter 9.13 and 14.6.

MOSFET CIRCUIT MODEL

All the work done in previuos cases comes together here in developing a circuit model of a MOSFET. There are many models possible; we have chosen here one that is well recognized in the literature and very useful for circuit analysis. Consult Figure 12.1 for the lumped component equivalent model and associated MOSFET parts. Each of the components may be calculated as described below.

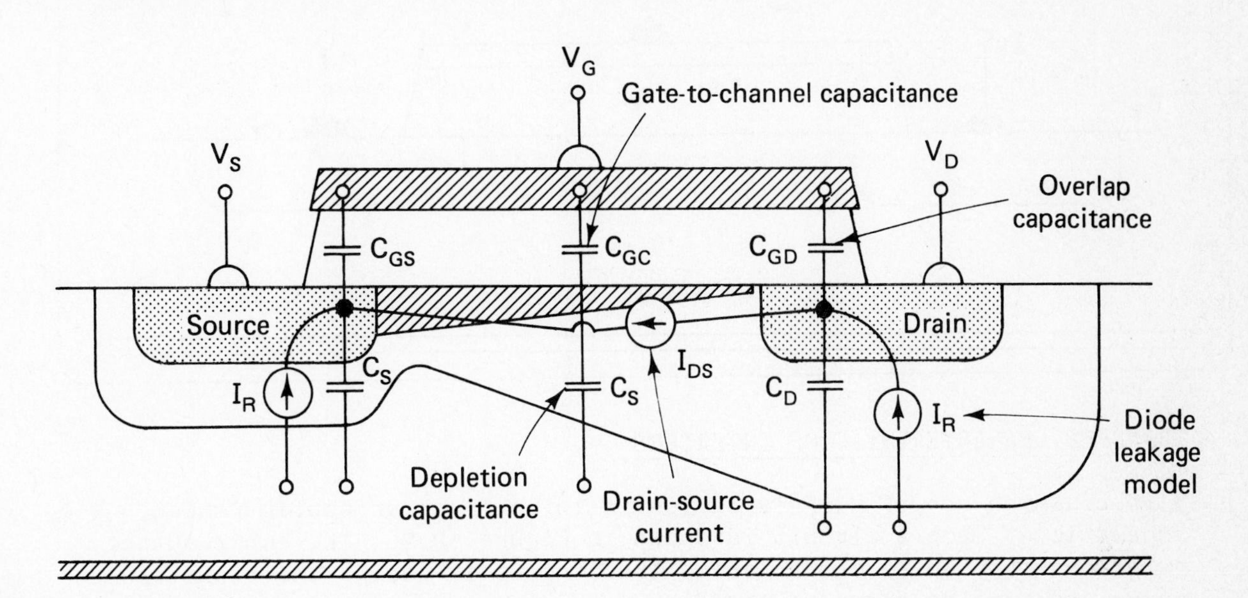

Figure 12.1 - Full MOSFET model employed in circuit analysis.

Source and Drain

I_R= diode current from source or drain junction to substrate. (See case 6).

C_{GS}, C_{GD}= gate to source and drain to source overlap capacitances, MOS parallel plate: gate to highly doped diffusion, commonly refered to as feedback parasitic capacitances. They are minimized by self-aligned processes. (See Case 3.)

Gate

C_{GG}= gate to channel capacitance, a very large input capacitance, ussually taken to be C_{MAX} for the gate. (See Case 3.)

Channel

C_{CS}= channel to substrate capacitance, a deep depletion like capacitance which may be modeled as in Case 4 for a given source to substrate bias.

I_{DS}= drain to source current, modeled by a dependent current source. (See Cases 5 and 11.)

Exercise 12.1

Consider the MOSFET described in Exercise 11.2. On the drawing below, properly label and give values for all material parameters for the MOSFET in question. Draw in and give expressions for all circuit elements with enough detail that it may easily be implemented in a computer code.

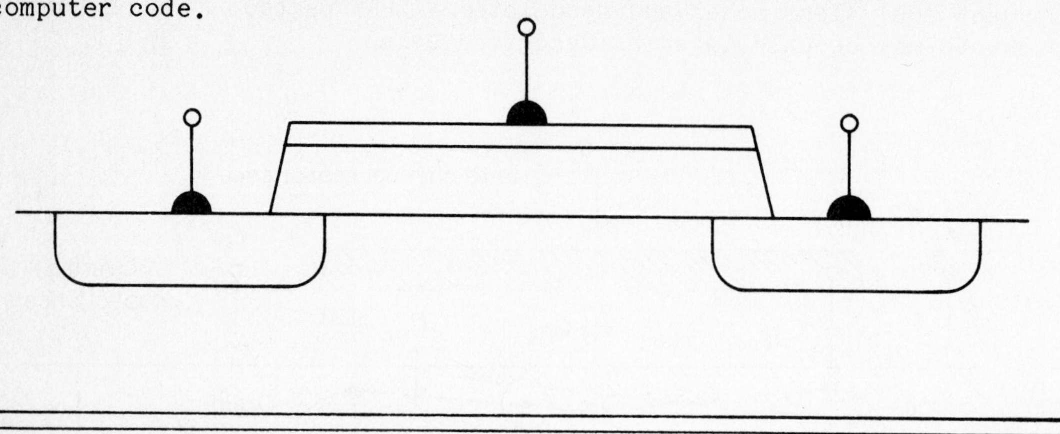

TRANSFER CHARACTERISTIC OF INVERTERS

A most common use of MOSFETs in ICs is in an inverter configuration. A schematic of such a circuit is shown in Figure 12.2. The input–output relationship used to characterize the circuit is called the transfer curve and is also shown in the figure.

Exercise 12.2

Using a 10 volt drain power supply, compute and plot the transfer characteristic of a resistive load inverter with two different loads, R_L=1 [Kohm] n and R_L=5 [Kohm]. Use the n channel MOSFET characteristic given in Figure 11.3 and employ the load line method. Use V_{SS}=0 volts and an input voltage swing of 10 volts.

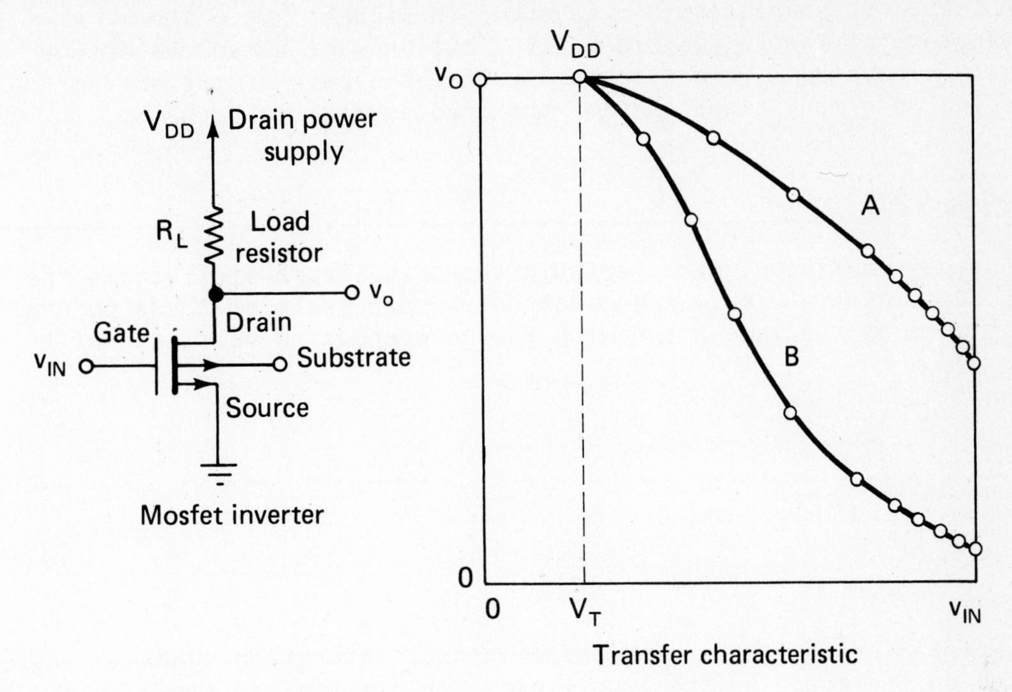

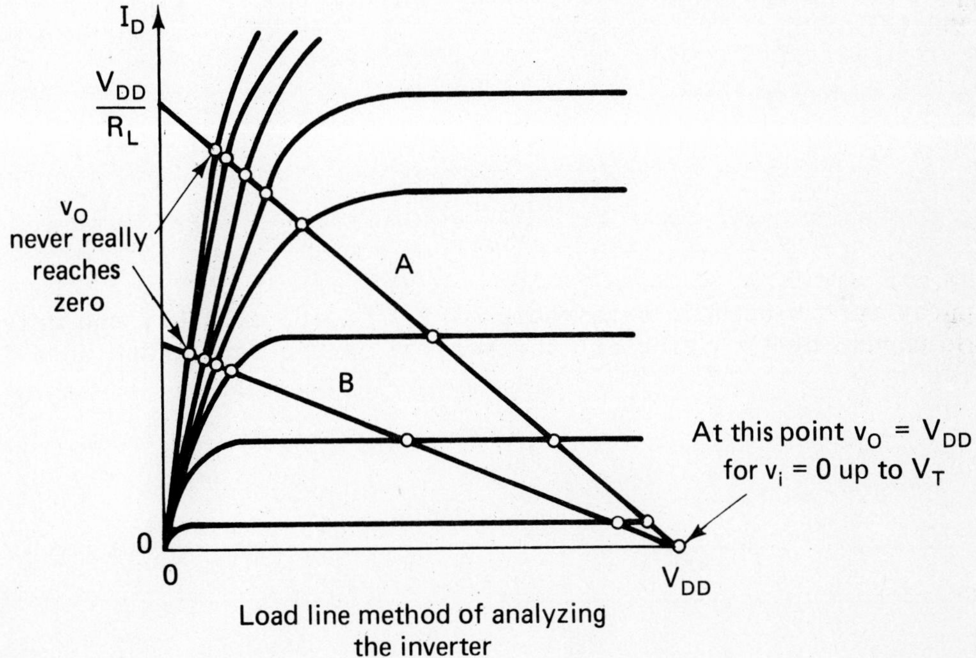

Figure 12.2 - Typical n channel resistive load inverter configuration, I_D - V_D characteristics and transfer curves.

Since integrated resistors take up too much space on the chip they are not used in advanced ICs. Another MOSFET is used in their place as the load device. If the gate of the load device is tied to its drain, then it is in saturation and one has essentially a resistor as a load with a parabolic I_D-V_D characteristic as shown in Figure 12.3. The disadvantage is that V_{oHIGH} is not V_{DD} but $V_{DD}-V_T$, a reduced output swing. To have V_{oLOW} close to zero volts, the beta of the driver device is made many times that of the load device, or $B_D/B_L > 1$.

Exercise 12.3

a. On the transistor characteristic shown in Figure 11.3 circle the locus of points $V_D = V_{SAT}$ and construct a data table of these points. Replot them on the same graph with the corresponding voltages shifted by $V_{DD}-V_T$.

b. Repeat Exercise 12.2 using the n channel saturation characteristic plotted on the graph as the load line. The results are for a beta ratio of 1:1.

c. Repeat part b using a beta ratio of 1:5 (i.e., multiply the driver device current by 5). Plot all the results on the same graph used in Exercise 12.2.

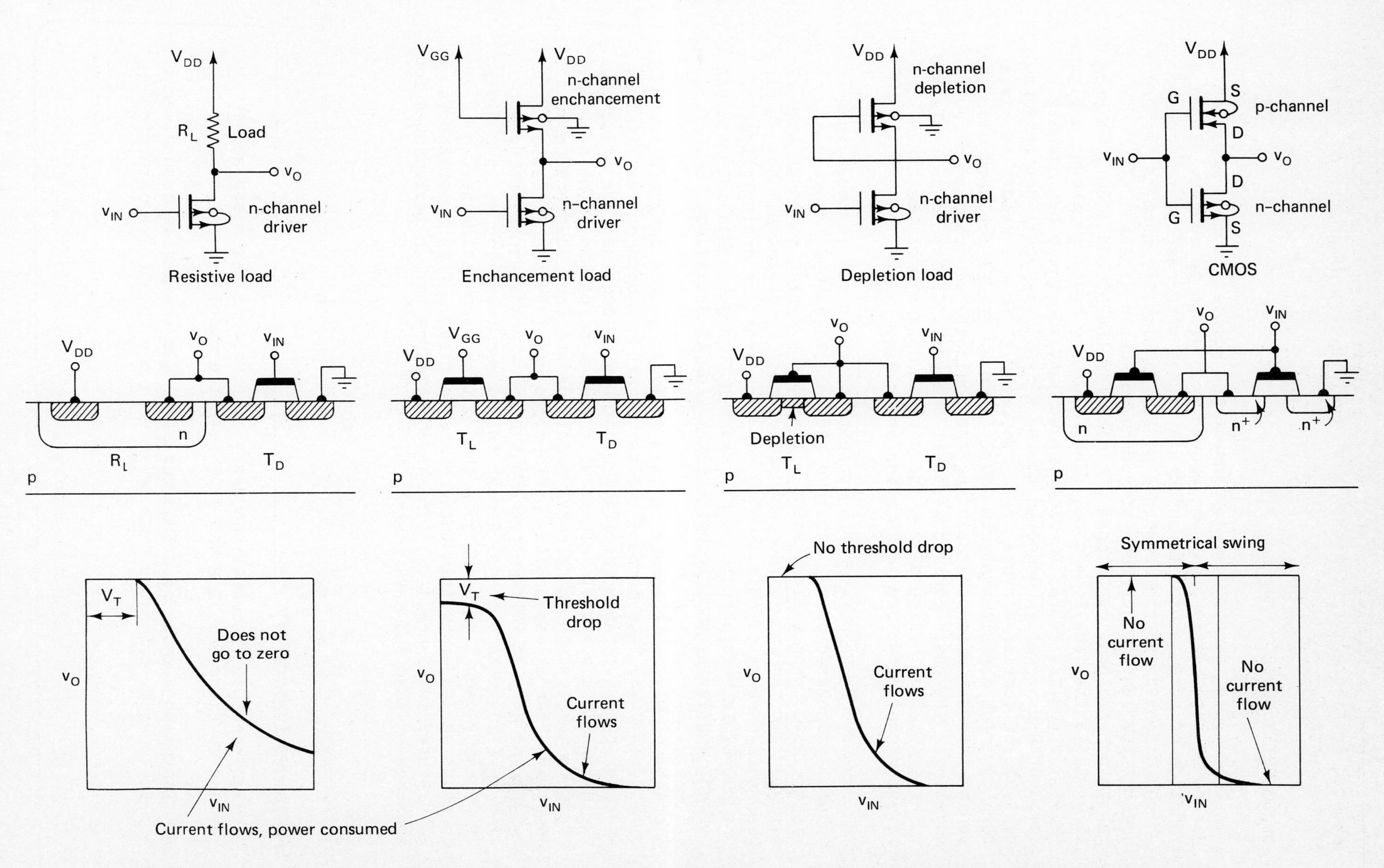

Figure 12.3 – Inverter configutaions: resistive load, non-saturated enhancement load, saturated enhancement load and CMOS. Circuit diagrams, device cross-sections and typical transfer curves are shown.

The transfer characteristics of various inverter configurations is plotted in Figure 12.3 together with their cicuit diagrams and device cross-sections.

DC COMPUTER MODEL

For DC modeling purposes we may reduce the complex circuit model of Figure 12.1 to simply a current source from drain to source. It is a function of gate, substrate and drain biases with respect to source, which is considered the reference node. To simplify matters, we may neglect the leakage currents of source and drain diodes. All capacitors are considered open circuits for the DC and slowly varying AC analysis that shall be done. The resulting model is shown in Figure 12.4a. To analyze MOSFET DC circuits we just add the other circuit components externally connected to the MOSFET. We shall not perform analysis of complex circuts here. We leave it to the reader to use a more powerful circuit analysis model (such as SPICE) for such circuits. We are interested in understanding the rudiments of computer circuit modeling. The basic circuit we shall analyze with the computer is the inverter. Thus in Figure 12.4 the circuit diagrams of the three inverter configurations in question are given.

The computer technique to obtain the transfer curves in these three cases is discussed below. Observe the resistive load inverter of Figure 12.4c. For a given input voltage, the current through the resistor and through the current source must be the same. Not knowing what this current is (since we do not know what the drain voltage of the MOSFET is beforehand), we can start with a guess and interate until current continuity and Kirchoff voltage law is satisfied.

A practical way of implementing this analysis on the computer is to start with $V_{DS} = V_{DD}$, compute the two currents, compare them; if they are not equal (within a specified error bound) decrease the value of V_{DS} by $V_{DD}/10$ and repeat. There will come a point where the two currents will change in relative magnitude and their difference will be negative where it was positive before. To continue the iteration, return one step in voltage decrement and now continue iterating but with $V_{DD}/100$ voltage steps. Repeat this inner iteration as many times as necesary each time decreasing the voltage step until the two currents are equal (within the error specified). This algorithm is illustrated in Figure 12.5. The same procedure may be implemented for enhancement load and CMOS inverters.

Computer Laboratory

1. Implement the DC model and analysis procedure described above and do Exercises 12.2 and 12.3 with the model.

2. Implement a model of a CMOS inverter. Use the device data given in Appendix C for the CMOS devices in the CD4007 RCA chip.

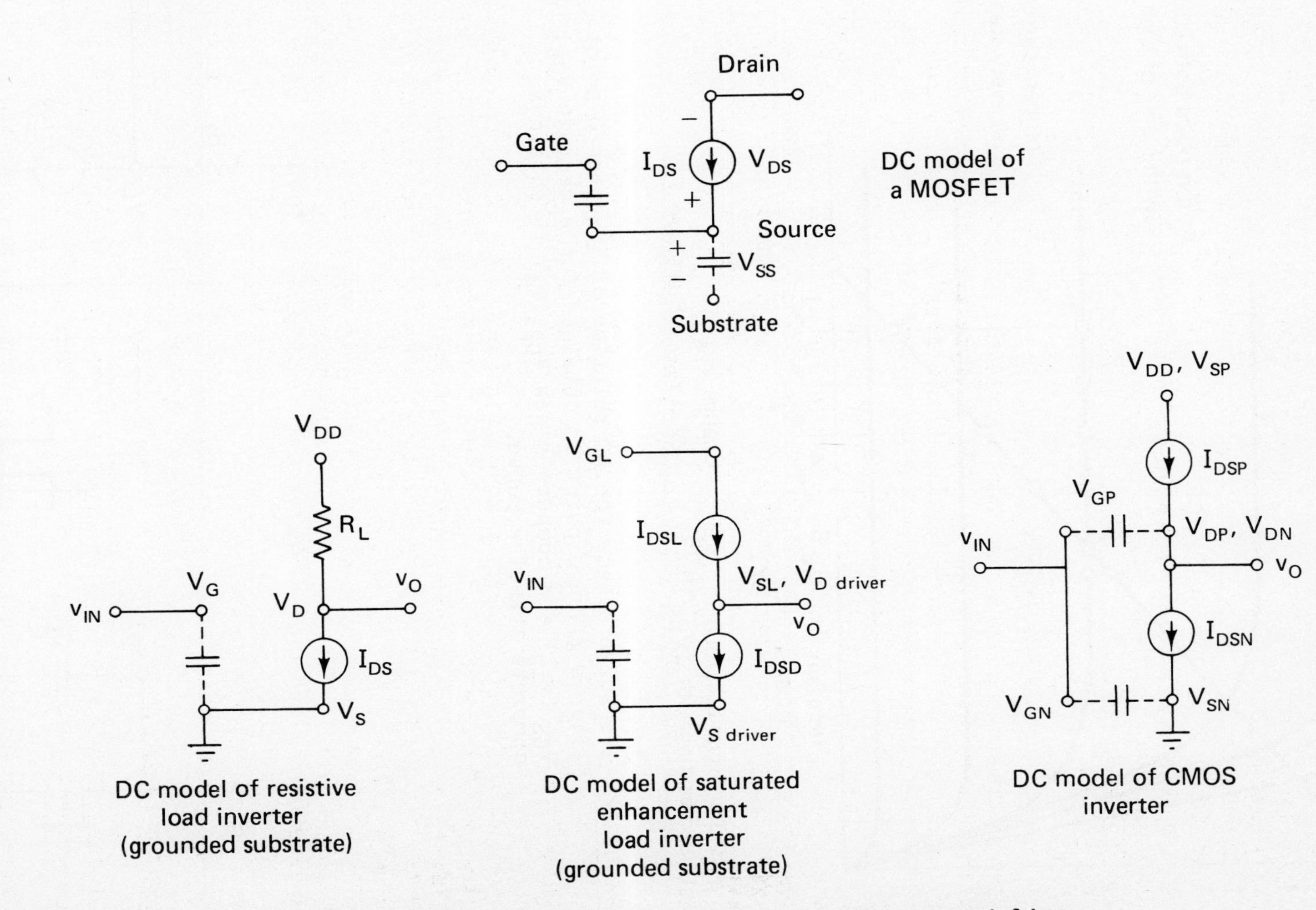

Figure 12.4 - Circuit models of various inverters useful for DC modeling.

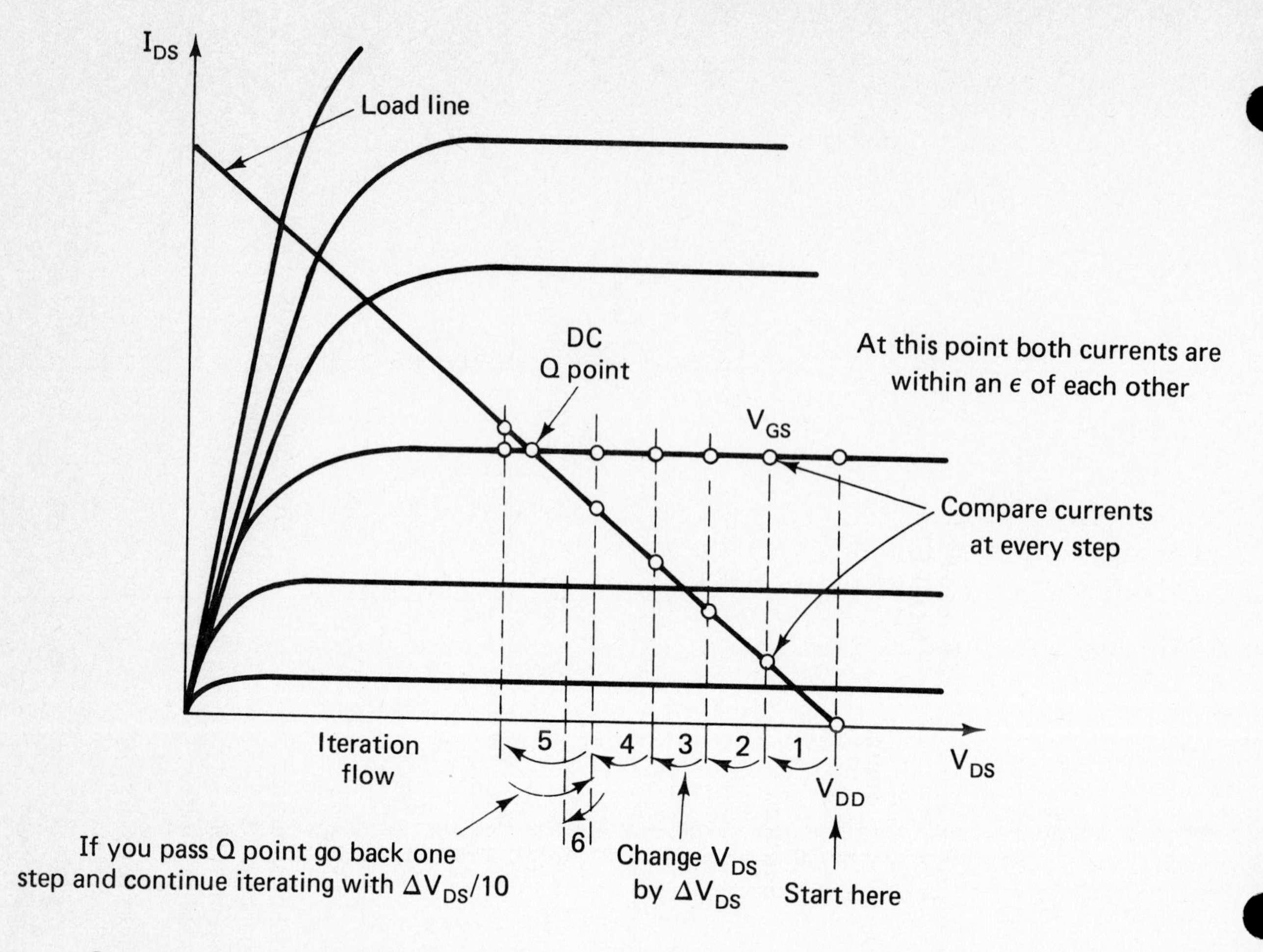

Figure 12.5 - Iterative scheme for DC solution of inverter Q point. After a given Q point is obtained for a certain V_{GS}, change V_{GS} (v_{in}) and compute a new V_{DS} (v_o) to obtain a complete transfer curve.

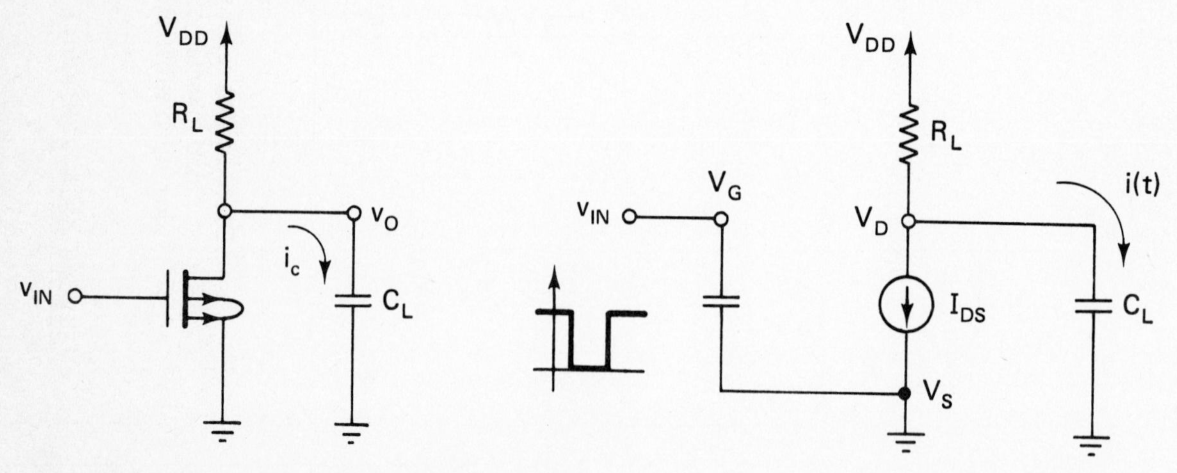

Figure 12.6 - Inverter circuit model with capacitive load useful for transient analysis.

TRANSIENT ANALYSIS OF INVERTERS

Inverters are often used in ICs as a buffer for a low current input signal, (such as a clock signal) to enable the addressing of the gates of many MOSFETs in parallel. Such a situation is encountered in Random Access Memories (RAMs). In this case the parallel gates are modeled as a large capacitor. The load device resistance and the load capacitance (the gates and associated conductors) make up the major portion of the RC time constant of the addressing circuit.

We see how important it is to perform the transient analysis of such a circuit. From this analysis we may make an estimate of the speed of operation of the chip! The turn on and turn off transient characteristics of this circuit are analyzed below.

In Figure 12.6 we have the schematic of a resistive load inverter charging a capacitor. A transient model of this circuit takes into account the current flowing through the resistor and current source. Here they are not equal but they differ by the charging current into the capacitor. Imposing Kirchoff's current laws at the capacitor resistor junction we obtain:

$$I_C = I_{RL} - I_{DS} = C\, dV_D/dt. \qquad (12.1)$$

Here we have made the approximation that the parallel gate capacitors may be modeled by a lumped fixed capacitance (use C_{MAX}). The differential of the drain voltage is approximated by a voltage difference between time increments. This problem then becomes similar to Case 7 where the incremental charging of a diode was analyzed.

The computer program for transient analysis makes use of the code already developed earlier in this case. Instead of the load device current and the driver current having to be equal for a solution, they must differ by the current going into the capacitor. The capacitor current is calculated using Equation 12.1, rewritten below:

$$I_C = C\, \Delta V_D / \Delta t. \qquad (12.2)$$

Of interest is the turn-on and turn-off time of the circuit. They are given as shown in Figure 12.7.

Computer Laboratory

Consider using all three inverters studied in Computer Laboratory A (Case 12) to charge capacitors. Let us consider also that 100 MOSFET gates in parallel are addressed with this circuit, (W/L=1, T_{ox}= 750Å). Build a computer model of this transient situation using the previous models as a starting point. Print out and plot V_{out} versus time from the time the input voltage is turned on to such a time that equlibrium is reached, that is: turned off. Use a 0-10 [volt] input voltage swing and 10 [volts] as a drain power supply. Estimate turn-on and turn-off times from your output.

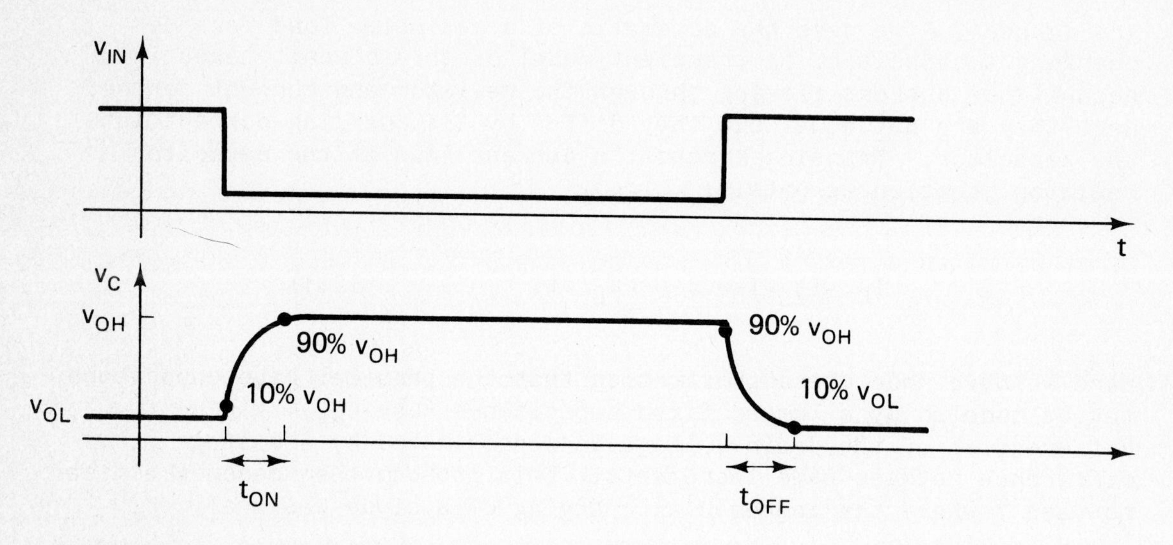

Figure 12.7 - Definitions of turn-on and turn-off for inverter circuit of Figure 12.6.

APPENDIX A

TABULATION OF THE COMPLEMENTARY ERROR FUNCTION

A tabulation of erfc(z) is given below. A discussion of this function is given in Cases 8, 9 and 10, where it is most often encountered. The algebraic approximation to the error function, useful for computer implementation, is also given below for convenience.

Z	erfc(Z)	Z	erfc(Z)	Z	erfc(Z)	Z	erfc(Z)	Z	erfc(Z)
0.00	1.00000	0.33	0.640721	0.66	0.350623	0.99	0.161492	1.32	0.619348D-01
0.01	0.988717	0.34	0.630635	0.67	0.343372	1.00	0.157299	1.33	0.599850D-01
0.02	0.977435	0.35	0.620618	0.68	0.336218	1.01	0.153190	1.34	0.580863D-01
0.03	0.966159	0.36	0.610670	0.69	0.329160	1.02	0.149162	1.35	0.562378D-01
0.04	0.954889	0.37	0.600794	0.70	0.322199	1.03	0.145216	1.36	0.544386D-01
0.05	0.943628	0.38	0.590990	0.71	0.315334	1.04	0.141350	1.37	0.526876D-01
0.06	0.932378	0.39	0.581261	0.72	0.308567	1.05	0.137564	1.38	0.509840D-01
0.07	0.921142	0.40	0.571608	0.73	0.301896	1.06	0.133856	1.39	0.493267D-01
0.08	0.909922	0.41	0.562031	0.74	0.295322	1.07	0.130227	1.40	0.477149D-01
0.09	0.898719	0.42	0.552532	0.75	0.288844	1.08	0.126674	1.41	0.461476D-01
0.10	0.887537	0.43	0.543113	0.76	0.282463	1.09	0.123197	1.42	0.446238D-01
0.11	0.876377	0.44	0.533775	0.77	0.276178	1.10	0.119795	1.43	0.431427D-01
0.12	0.865242	0.45	0.524518	0.78	0.269990	1.11	0.116467	1.44	0.417034D-01
0.13	0.854133	0.46	0.515345	0.79	0.263897	1.12	0.113212	1.45	0.403050D-01
0.14	0.843053	0.47	0.506255	0.80	0.257899	1.13	0.110029	1.46	0.389465D-01
0.15	0.832004	0.48	0.497250	0.81	0.251997	1.14	0.106918	1.47	0.376271D-01
0.16	0.820988	0.49	0.488332	0.82	0.246189	1.15	0.103876	1.48	0.363459D-01
0.17	0.810008	0.50	0.479500	0.83	0.240476	1.16	0.100904	1.49	0.351021D-01
0.18	0.799064	0.51	0.470756	0.84	0.234857	1.17	0.979996D-01	1.50	0.338949D-01
0.19	0.788160	0.52	0.462101	0.85	0.229332	1.18	0.951626D-01	1.51	0.327233D-01
0.20	0.777297	0.53	0.453536	0.86	0.223900	1.19	0.923917D-01	1.52	0.315865D-01
0.21	0.766478	0.54	0.445061	0.87	0.218560	1.20	0.896860D-01	1.53	0.304838D-01
0.22	0.755704	0.55	0.436677	0.88	0.213313	1.21	0.870445D-01	1.54	0.294143D-01
0.23	0.744977	0.56	0.428384	0.89	0.208157	1.22	0.844661D-01	1.55	0.283773D-01
0.24	0.734300	0.57	0.420184	0.90	0.203092	1.23	0.819499D-01	1.56	0.273719D-01
0.25	0.723674	0.58	0.412077	0.91	0.198117	1.24	0.794948D-01	1.57	0.263974D-01
0.26	0.713100	0.59	0.404063	0.92	0.193232	1.25	0.770999D-01	1.58	0.254530D-01
0.27	0.702582	0.60	0.396144	0.93	0.188436	1.26	0.747640D-01	1.59	0.245380D-01
0.28	0.692120	0.61	0.388319	0.94	0.183729	1.27	0.724864D-01	1.60	0.236516D-01
0.29	0.681716	0.62	0.380589	0.95	0.179109	1.28	0.702658D-01	1.61	0.227932D-01
0.30	0.671373	0.63	0.372954	0.96	0.174576	1.29	0.681014D-01	1.62	0.219619D-01
0.31	0.661092	0.64	0.365414	0.97	0.170130	1.30	0.659920D-01	1.63	0.211572D-01
0.32	0.650874	0.65	0.357971	0.98	0.165768	1.31	0.639369D-01	1.64	0.203782D-01
								1.65	0.196244D-01

Z	erfc(Z)
1.66	0.188951D-01
1.67	0.181896D-01
1.68	0.175072D-01
1.69	0.168474D-01
1.70	0.162095D-01
1.71	0.155930D-01
1.72	0.149972D-01
1.73	0.144215D-01
1.74	0.138654D-01
1.75	0.133283D-01
1.76	0.128097D-01
1.77	0.123091D-01
1.78	0.118258D-01
1.79	0.113594D-01
1.80	0.109095D-01
1.81	0.104755D-01
1.82	0.100568D-01
1.83	0.965319D-02
1.84	0.926405D-02
1.85	0.888897D-02
1.86	0.852751D-02
1.87	0.817925D-02
1.88	0.784378D-02
1.89	0.752068D-02
1.90	0.720957D-02
1.91	0.691006D-02
1.92	0.662177D-02
1.93	0.634435D-02
1.94	0.607743D-02
1.95	0.582066D-02
1.96	0.557372D-02
1.97	0.533627D-02
1.98	0.510800D-02
1.99	0.488859D-02
2.00	0.467773D-02
2.01	0.447515D-02
2.02	0.428055D-02
2.03	0.409365D-02
2.04	0.391419D-02
2.05	0.374190D-02
2.06	0.357654D-02
2.07	0.341785D-02
2.08	0.326559D-02
2.09	0.311954D-02
2.10	0.297947D-02
2.11	0.284515D-02
2.12	0.271639D-02
2.13	0.259298D-02
2.14	0.247471D-02
2.15	0.236139D-02
2.16	0.225285D-02
2.17	0.214889D-02
2.18	0.204935D-02
2.19	0.195406D-02
2.20	0.186285D-02
2.21	0.177556D-02
2.22	0.169205D-02
2.23	0.161217D-02
2.24	0.153577D-02
2.25	0.146272D-02
2.26	0.139288D-02
2.27	0.132613D-02
2.28	0.126234D-02
2.29	0.120139D-02
2.30	0.114318D-02
2.31	0.108758D-02
2.32	0.103449D-02
2.33	0.983805D-03
2.34	0.935430D-03
2.35	0.889267D-03
2.36	0.845223D-03
2.37	0.803210D-03
2.38	0.763142D-03
2.39	0.724936D-03
2.40	0.688514D-03
2.41	0.653798D-03
2.42	0.620716D-03
2.43	0.589197D-03
2.44	0.559174D-03
2.45	0.530580D-03
2.46	0.503353D-03
2.47	0.477434D-03
2.48	0.452764D-03
2.49	0.429288D-03
2.50	0.406952D-03
2.51	0.385705D-03
2.52	0.365499D-03
2.53	0.346286D-03
2.54	0.328021D-03
2.55	0.310660D-03
2.56	0.294163D-03
2.57	0.278489D-03
2.58	0.263600D-03
2.59	0.249461D-03
2.60	0.236034D-03
2.61	0.223289D-03
2.62	0.211191D-03
2.63	0.199711D-03
2.64	0.188819D103
2.65	0.178488D-03
2.66	0.168689D-03
2.67	0.159399D-03
2.68	0.150591D-03
2.69	0.142243D-03
2.70	0.134333D-03
2.71	0.126838D-03
2.72	0.119738D-03
2.73	0.113015D-03
2.74	0.106649D-03
2.75	0.100622D-03
2.76	0.949176D-04
2.77	0.895197D-04
2.78	0.844127D-04
2.79	0.795818D-04
2.80	0.750132D-04
2.81	0.706933D-04
2.82	0.666096D-04
2.83	0.627497D-04
2.84	0.591023D-04
2.85	0.556563D-04
2.86	0.524012D-04
2.87	0.493270D-04
2.88	0.464244D-04
2.89	0.436842D-04
2.90	0.410979D-04
2.91	0.386573D-04
2.92	0.363547D-04
2.93	0.341828D-04
2.94	0.321344D-04
2.95	0.302030D-04
2.96	0.283823D-04
2.97	0.266662D-04
2.98	0.250491D-04
2.99	0.235256D-04
3.00	0.220905D-04
3.01	0.207390D-04
3.02	0.194664D0-4
3.03	0.182684D-04
3.04	0.171409D-04
3.05	0.160798D-04
3.06	0.150816D-04
3.07	0.141426D-04
3.08	0.132595D-04
3.09	0.124292D-04
3.10	0.116487D-04
3.11	0.109150D-04
3-12	0.102256D-04
3-13	0.957795D-05
3.14	0.896956D-05
3.15	0.839821D-05
3.16	0.786174D-05
3.17	0.735813D-05
3.18	0.688545D-05
3.19	0.644190D-05
3.20	0.602576D-05
3.21	0.563542D-05
3.22	0.526935D-05
3.23	0.492612D-05
3.24	0.460435D-05
3.25	0.430278D-05
3.26	0.402018D-05
3.27	0.375542D-05
3.28	0.350742D-05
3.29	0.327517D-05
3.30	0.305771D-05
3.31	0.285414D-05
3.32	0.266360D-05
3.33	0.248531D-05
3.34	0.231850D-05
3.35	0.216248D-05
3.36	0.201656D-05
3.37	0.188013D-05
3.38	0.175259D-05
3.39	0.163338D-05
3.40	0.152199D-05
3.41	0.141793D-05
3.42	0.132072D-05
3.43	0.122994D-05
3.44	0.114518D-05
3.45	0.106605D-05
3.46	0.992201D-06
3.47	0.923288D-06
3.48	0.858995D-06
3.49	0.799025D-06
3.50	0.743098D-06
3.51	0.690952D-06
3.52	0.642341D-06
3.53	0.597035D-06
3.54	0.554816D-06
3.55	0.515484D-06
3.56	0.478847D-06
3.57	0.444728D-06
3.58	0.412960D-06
3.59	0.383387D-06
3.60	0.355863D-06
3.61	0.330251D-06
3.62	0.306423D-06
3.63	0.284259D-06
3.64	0.263647D-06
3.65	0.244483D-06
3.66	0.226667D-06
3.67	0.210109D-06
3.68	0.194723D-06
3.69	0.180429D-06
3.70	0.167151D-06
3.71	0.154821D-06
3.72	0.143372D-06
3.73	0.132744D-06
3.74	0.122880D-06
3.75	0.113727D-06
3.76	0.105236D-06
3.77	0.973591D-07
3.78	0.900547D-07
3.79	0.832821D-07
3.80	0.770039D-07
3.81	0.711851D-07
3.82	0.657933D-07
3.83	0.607981D-07
3.84	0.561711D-07
3.85	0.518863D-07
3.86	0.479189D-07
3.87	0.442464D-07
3.88	0.408473D-07
3.89	0.377021D-07
3.90	0.347922D-07
3.91	0.321007D-07
3.92	0.296117D-07
3.93	0.273103D-07
3.94	0.251829D-07
3.95	0.232167D-07
3.96	0.213999D-07
3.97	0.197214D-07
3.98	0.181710D-07
3.99	0.167392D-07
4.00	0.154173D-07
4.01	0.141969D-07
4.02	0.130707D-07
4.03	0.120314D-07
4.04	0.110726D-07
4.05	0.101882D-07
4.06	0.937269D-08
4.07	0.862073D-08
4.08	0.792756D-08
4.09	0.728870D-08
4.10	0.670003D-08
4.11	0.615769D-08
4.12	0.565816D-08
4.13	0.519813D-08
4.14	0.477457D-08
4.15	0.438468D-08
4.16	0.402583D-08
4.17	0.369564D-08
4.18	0.339186D-08
4.19	0.311245D-08
4.20	0.285549D-08
4.21	0.261924D-08
4.22	0.240207D-08
4.23	0.220247D-08
4.24	0.201907D-08
4.25	0.185057D-08
4.26	0.169581D-08
4.27	0.155369D-08
4.28	0.142319D-08
4.29	0.130341D-08
4.30	0.119347D-08

USEFUL APPROXIMATION TO THE ERROR FUNCTION

For computer implementation of the error function the following rational approximation is given by Abromowitz and Stegun in Handbook of Mathematical Functions, Dover, N.Y., page 299:

$$erf(x) = 1 - (a_1 t + a_2 t^2 + a_3 t^3)\ e^{-x^2}$$

where $t = 1/(1+px)$, and

$p = .47047$, $a_1 = .3480242$, $a_2 = -.0958798$, $a_3 = .7478556$.

APPENDIX B

STATISTICS FOR ION IMPLANTED DOPANTS

PROJECTED RANGES FOR COMMON DOPANTS

BORON IN SILICON			PHOSPHORUS IN SILICON			ARSENIC IN SILICON		
Energy (KEV)	Projected Range (Microns)	Projected Standard Deviation (Microns)	Energy (KEV)	Projected Range (Microns)	Projected Standard Deviation (Microns)	Energy (KEV)	Projected Range (Microns)	Projected Standard Deviation (Microns)
10	0.0344	0.0156	10	0.0144	0.0070	10	0.0096	0.0037
20	0.0674	0.0253	20	0.0260	0.0122	20	0.0159	0.0060
30	0.0999	0.0331	30	0.0375	0.0169	30	0.0216	0.0081
40	0.1311	0.0392	40	0.0490	0.0214	40	0.0271	0.0101
50	0.1616	0.0445	50	0.0610	0.0259	50	0.0324	0.0120
60	0.1914	0.0491	60	0.0732	0.0303	60	0.0377	0.0139
70	0.2202	0.0530	70	0.0855	0.0345	70	0.0429	0.0158
80	0.2478	0.0564	80	0.0980	0.0386	80	0.0481	0.0176
90	0.2739	0.0593	90	0.1106	0.0425	90	0.0533	0.0194
100	0.2992	0.0618	100	0.1233	0.0463	100	0.0584	0.0211
110	0.3238	0.0641	110	0.1362	0.0500	110	0.0635	0.0229
120	0.3479	0.0663	120	0.1491	0.0535	120	0.0686	0.0246
130	0.3714	0.0682	130	0.1621	0.0570	130	0.0738	0.0263
140	0.3942	0.0700	140	0.1752	0.0604	140	0.0789	0.0279
150	0.4165	0.0716	150	0.1883	0.0636	150	0.0840	0.0296
160	0.4383	0.0731	160	0.2014	0.0668	160	0.0891	0.0312
170	0.4594	0.0744	170	0.2146	0.0698	170	0.0943	0.0329
180	0.4799	0.0757	180	0.2277	0.0726	180	0.0995	0.0345
190	0.5000	0.0769	190	0.2407	0.0753	190	0.1048	0.0362
200	0.5198	0.0780	200	0.2538	0.0780	200	0.1101	0.0378

APPENDIX C

LABORATORY EXPERIMENT 1

OBJECTIVES: (In reference to MOSFETS)

1. To measure the substrate sensitivity characteristics;
2. To measure the drain current–drain voltage characteristics;
3. To measure the channel mobility.

REFERENCE:

G.J. Herskowitz and R.B. Schilling, Semiconductor Device Modeling for Computer-Aided Design, McGraw Hill, N.Y., 1972,(see Chapter 5).

DEVICE: RCA CD4007 Dual Complementary Pair Plus Inverter (see data sheets).

EQUIPMENT:

DC power supplies
Oscilloscope
DC voltmeter
Function generator
Prototype board
DC microammeter

DEVICE PATRAMETER DATA: CD4007

N channel	P channel
$W = 7.5$ [mils];	$W = 19.5$ [mils];
$L = .16$ [mils];	$L = .25$ [mils];
$t_{ox} = 1200\text{Å}$;	$t_{ox} = 1200\text{Å}$;
$\mu = 420$ [cm^2/V sec];	$\mu = 165$ [cm^2/V sec];
$N_A = 2.8 \times 10^{16}$ [cm^{-3}];	$N_D = 3 \times 10^{15}$ [cm^{-3}].

1. SUBSTRATE SENSITIVITY

a. Consult the data sheets and BUILD the circuit shown in Figure C.1 with an n channel MOSFET.

b. CHOOSE a value of V_{DS} to bias the device well into saturation for the entire range of gate biases (0-10 [volts]). Set V_{SS}=0.0 [volts].

c. Vary the gate bias until sufficient data (measure I_{DS}) has been MEASURED to accurately plot (I_{DS}) versus V_{GS}. REPEAT using at least six more different substrate biases (2,4,6,8,10 and 12 [volts]).

d. PLOT (IDS) vs VGS for all seven sets of data on the same graph. Extrapolate and determine the thresholds by the method shown in Case 11 of the Workbook.

e. PLOT threshold versus substrate bias as a substrate sensitivity characteristic.

f. Use the device parameters given and SIMULATE the substrate sensitivity with the model developed in Case 5. PLOT on the same graph as (e) above, using zero flatband. ESTIMATE the flatband voltage and Q_{SS}.

g. REPEAT parts a-f for a p channel enhancement device on the same chip.

2. DRAIN CURRENT VERSUS DRAIN VOLTAGE CHARACTERISTIC

a. Using the circuit shown in Figure C.1 but now with the substrate grounded MEASURE drain current as a function of drain bias (0-15 [volts]) in 0.5 [volt] increments for various gate biases (0-15 [volts] in 1 [volt] steps).

b. PLOT this set of characteristics as they shall be most important for design purposes in future experiments.

c. Use a curve tracer and GENERATE this family of curves. OBTAIN a photograph of this trace if possible and compare to the results of (b) above.

d. Use the device data given and SIMULATE these characteristics using the models developed in Case 11.

e. COMPUTE the mobility of minority carriers for this device using the saturation current characteristics by the technique shown in Case 11. PLOT it versus gate voltage and COMPARE to the average mobility, the given mobility and the expected bulk mobility for this type material

f. REPEAT parts a-e for a p channel enhancement MOSFET device on the same chip.

3. CHANNEL MOBILITY

COMPUTE the channel mobility of the n channel MOSFET as well as the p channel MOSFET from the measurement of drain current in saturation (see Part 2.) PLOT mobility for both devices as a function of gate bias–threshold. EXPLAIN any variations in mobility.

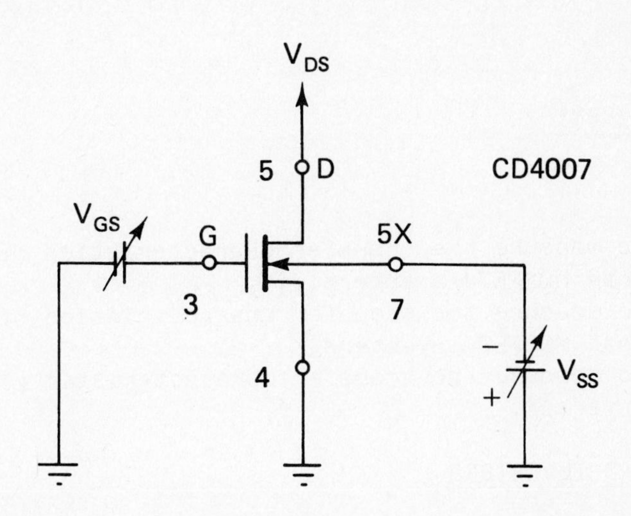

Figure C.1 – MOSFET test circuit to obtain a substrate sensitivity curve. The MOSFET being tested is an n channel device on the RCA CD4007 chip. For a p channel device, reverse the polarity on the bias power supplies.

LABORATORY EXPERIMENT 2

OBJECTIVES:

1. To measure the transfer characteristics of resistive load MOSFET inverters.
2. To measure the transfer characteristics of enhancement load MOSFET inverters.
3. To measure the transfer chraracteristic of a CMOS inverter.

1. RESISTIVE LOAD INVERTERS

a. WIRE the inverter circuit shown (Figure C.2) and MEASURE the transfer characteristics (use a 0-10 [volts] input voltage swing) for the grounded substrate n channel MOSFET circuit. Use the CD4007 chip and a 1 [Kohm] and 500 [ohm] resistors for the loads.

b. USE the model of Case 12 to predict these two transfer curves. (If Case 12 was not completed use the load line method and the data taken from Experiment 1, part 2 to predict the measured transfer curves.)

c. DRAW both predicted and measured curves on the same graph.

d. DESIGN a resistive load n channel inverter for a 0-10 [volt] input voltage swing and a V_{out} (high)=10 [v]. and V_{out} (low)= 1 [v].. (Use the CD4007 devices).

2. ENHANCEMENT LOAD INVERTERS

a. WIRE the n channel MOSFET saturated enhancement load circuit shown in Figure C.3. Use two n channel MOSFETS from the CD4007 chip for a 1:1 beta ratio and repeat for a 2:1 beta ratio by wiring two n channel devices in parallel as the driver.

b. MEASURE the transfer characteristics (use a 0-10 [volt] input voltage swing) for both configurations.

c. COMPUTE a predicted tranfer curve from the model of Case 12 or from the data obtained in Experiment 1, part 2.

d. DRAW both data and model on the same graph and COMPARE.

3. CMOS INVERTERS

a. WIRE the CMOS inverter shown in Figure C.3 and MEASURE the transfer characteristics. Use the CD4007 devices and a 0-10 [volt] input voltage swing.

b. COMPARE the results with the published data from the manufacturer. DRAW both on the same graph.

4. COMPARISON

In conclusion, COMPARE all three transfer curves and make a statement of your conclusions and observations compared to expectations.

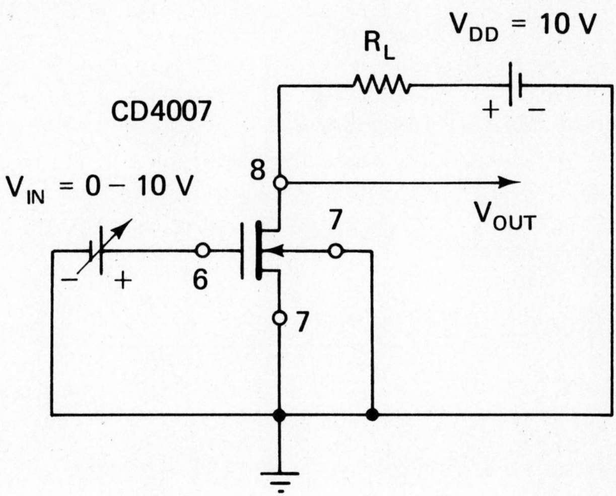

Figure C.2 - MOSFET test circuit used to obtain the transfer characteristics of a resistive load inverter.

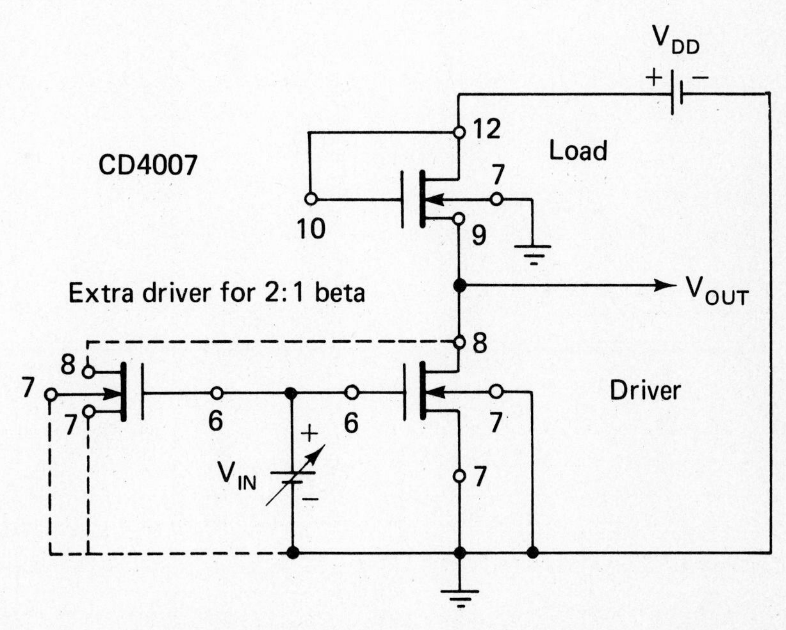

Figure C.3 - MOSFET test circuit used to obtain the transfer characteristics of a saturated load inverter. A 1:1 beta configuration is shown, as well as how to wire an extra device in parallel for a 2:1 beta ratio.